Genetic Engineering 4

To a good student, with the wishing that he'll continue on the same course.

6/4/'85

Carlo Bruschi

Genetic Engineering 4

Edited by

Robert Williamson

Professor of Biochemistry,
St Mary's Hospital Medical School,
University of London

ACADEMIC PRESS · 1983

A Subsidiary of Harcourt Brace Jovanovich, Publishers

London · New York
Paris · San Diego · San Francisco
São Paulo · Sydney · Tokyo · Toronto

ACADEMIC PRESS INC. (LONDON) LTD
24/28 Oval Road,
London NW1

United States Edition published by
ACADEMIC PRESS INC.
111 Fifth Avenue,
New York, New York 10003

British Library Cataloguing in Publication Data

Genetic engineering. — Vol. 4
1. Genetic engineering
575.1 QH442

ISBN 0–12–270304–9

LCCCN 80–41976

Printed in Great Britain at the Alden Press
Oxford London and Northampton

Contributors

R. K. Craig *Courtauld Institute of Biochemistry, The Middlesex Hospital Medical School, Mortimer Street, London W1P 7PN, UK*

R. Everett *MRC Virology Unit, Institute of Virology, Church Street, Glasgow G11 5JR, UK*

L. Hall *Courtauld Institute of Biochemistry, The Middlesex Hospital Medical School, Mortimer Street, London W1P 7PN, UK*

T. Harris *Celltech Limited, 244—250 Bath Road, Slough SL2 4DY, UK*

R. F. Lathe *Transgenè S. A., 11 rue Humann, 67000 Strasbourg, France*

J. P. Lecocq *Transgenè S. A., 11 rue Humann, 67000 Strasbourg, France*

Preface

In the two years since the first volume of this series was conceived, a remarkable change has taken place, one of which few of us foresaw. From a research technique mastered by a few hundred scientists, recombinant DNA is now a tool of virtually every contemporary laboratory in any field of biology or biochemistry: a routine method for protein sequencing, the way to determine gene number, domain evolution, chromosome localisation or family inheritance. During the next five years, it will probably become the standard technique for antenatal diagnosis, for crop improvement and for the industrial production of many chemicals, not only pharmaceuticals but also heavy chemicals and foodstuffs.

We see an absurd claim from time to time in the financial pages of the press: "the genetic engineering bubble has burst". I honestly wonder what some journalists spend their time doing! Having hyped the biotechnology field for the sake of "exciting copy", and succeeded in overestimating the potential in the *short run* of even this limitless field, they now burst their own bubble and pretend that the field is going down. The truth is that there is an enormous potential in biotechnology for research and application, but this will come to fruition over the next decade, not in the next three months. The payoff for the community, in any case, will not be measured in dollars or pounds, but in the contributions that genetic engineering makes to pure and applied clinical, agricultural, biological and industrial science.

My first Preface two years ago expressed the hope that the initial four volumes of "Genetic Engineering" would combine to give a complete primer of recombinant DNA technology. Thanks to the contributors, I think that the series comes close to achieving this object. We have covered phage, plasmid and virus cloning systems and discussed yeast, bacteria and animal cells, and looked at clinical and industrial applications as well as many fundamental uses to which genetic engineering can be put. To the extent that we have failed, the fault is mine as editor. It would have been good to have had articles about plant systems, and the industrial and clinical applications have come about faster than even I, an incurable optomist, thought possible. I understand that Peter Rigby, who is taking over as editor after this volume, means to correct some of these weaknesses as quickly as possible.

In this volume, Len Hall and Roger Craig discuss the most recent results obtained studying polypeptide hormones, and succeed once again in introducing a great deal of up-to-the-minute data, as have all the contributors. Rick Lathe, Dick Everett and Jean-Pierre Lecocq provide a detailed review of chemical techniques for site-directed mutagenesis and manipulation of restriction sites, an article which I feel will be a fine manual for anyone starting work on the chemical side of genetic engineering. Finally, Tim Harris has written an up-to-date summary of the way in which industry is seeking to obtain the expression of cloned genes at a high level, extremely timely at the moment when both human insulin and interferon synthesised using cloned genes are undergoing clinical trials.

I found it a pleasure to edit these volumes, largely because of the commitment of the authors and the help I have received from the staff of Academic Press. I hope that the book has been useful. I commented in the first preface that the only justification for a series such as this is the fact that people find it of value for teaching and for research, and if it meets that need, I will be pleased.

London, November 1982 *Bob Williamson*

Contents

Recombinant DNA technology: application to the characterisation and expression of polypeptide hormones

R. K. Craig and L. Hall

Expression of eukaryotic genes in *E. coli*

T. J. R. Harris

DNA engineering: the use of enzymes, chemicals and oligonucleotides to restructure DNA sequences *in vitro*

R. F. LATHE and J.P. LECOCQ

Transgène S.A., 11 rue Humann, 67000 Strasbourg, France

R. EVERETT

Laboratoire de Génètique Molèculaire du CNRS, 11 rue Humann, 67085 Strasbourg, France. Present address: MRC Unit of Virology, Church Street, Glasgow, UK

GENETIC ENGINEERING
ISBN 0 12 270304 9

I Introduction

It is only during the past few years that scientists have been able to manipulate DNA in any precise manner. DNA engineering is a new technology, using methods and techniques often quite dissimilar from those of classical biochemistry. The enormous amplification of a replicating DNA segment occurring after introduction into a host cell has changed the character of DNA analysis, such that primary manipulations are now performed upon minute quantities of material. Indeed, some experimental protocols involve manipulation of a population comprising only several thousand molecules.

In the present review we discuss the advantages and limitations of certain central techniques of DNA manipulation, and we aim primarily at the clarification of technical and theoretical details. It will be noted that DNA engineering is used by two quite different schools. The fundamental scientist is more concerned with altering a DNA sequence in order to determine its function, whereas the applied scientist may be interested in manipulating genes coding for products of medical or commercial importance. In practice, both groups share the core of common techniques which form the subject of this review.

In the first sections we discuss the tools used for DNA engineering: the restriction enzymes, ligases, polymerases and nucleases which make the manipulation of DNA possible. We stress their properties and have indicated how they have been or could be used in a sometimes quite complex manner to achieve a desired result. We go on to describe the uses of small synthetic oligonucleotides which have met increasing application in this field for they permit manipulations which involve successive transfers of DNA segments from one genetic location to another. The section on methylation of DNA illustrates how unexpected problems may be encountered during routine manipulation, and some ways of circumventing them. Finally we describe the application of these methods to the mutagenesis of DNA sequences, in particular to the alteration of a sequence lying at some distance from landmarks such as restriction enzyme sites. This approach has led in recent years to a considerable increase in our understanding of gene structure and function.

Throughout we have avoided the temptation to include a historical survey of the literature, nor do we describe experiments in the same detail as would be found in a laboratory manual. For these reasons certain older techniques have been omitted as they have been superseded by more efficient procedures. The principles, applications and limitations of techniques are presented in the hope that the reader may be able to apply these methods to any experimental problem in this field.

A Abbreviations and definitions

All nucleotide sequences are presented $5' \rightarrow 3'$ except where otherwise specified. Positions of cleavage within recognition sequences are represented by /, e.g. G/AATTC. Where the recognition sequence and cleavage site are symmetrical, only one strand is presented. ds = double strand; ss = single strand; N = any nucleotide; dNTP = any of the 4 deoxyribonucleoside triphosphates; bp = base pair(s): kb = kilobase pair(s), EGTA = ethylene glycol-bis (β-amino ethyl ether)-N, N′-tetra acetic acid; PPi = inorganic pyrophosphate; NAD = nicotinamide adenine mononucleotide; NMN = nicotinamide mononucleotide.

II Breaking and joining DNA

In this section we discuss the basic principles and materials used to manipulate DNA. These are the restriction enzymes and ligases which allow site-specific cleavage of DNA and *in vitro* ligation.

A Type II restriction endonucleases

This topic has been reviewed earlier in this Series (Malcolm, 1981) and elsewhere (Modrich, 1979; Wells *et al.*, 1981) and we will restrict our discussion accordingly. In contrast to type I and type III restriction endonucleases those of class II cleave ds DNA at specific sites (for a discussion of types I and III, see Yuan, 1981). Type II restriction endonucleases recognise a sequence of bases in DNA, usually 4 to 6 in length, and cleave ds DNA at a defined position in both strands at or near the recognition sequence. Roberts (1981) presents a detailed listing of restriction enzymes, their recognition sequences and exact cleavage sites; we will mention a few examples from this list to illustrate specific points. Type III enzymes could potentially be used for specific cleavage of DNA since, unlike the type I enzymes which generally cleave at a considerable distance from their recognition sites, the cleavage sites of type III enzymes fall within a restricted number of base pairs in the proximity of the recognition site.

We will nevertheless concentrate our discussion on restriction enzymes of type II: in the body of this article we will use restriction enzyme to refer exclusively to those of the second class. The actual position of cleavage in each strand relative to the recognition sequence varies from one enzyme to the next, and the enzymes can be subclassified on this basis.

Enzymes generating only one type of terminus: enzymes of this class recognise a simple, usually rotationally symmetrical, sequence of bases and cleave within this sequence. Examples are *Eco*RI and *Pst*I which cleave as below:

$$\begin{array}{llcllr} \textit{Eco}\text{RI} & 5'\text{--GAATTC--}3' & & \text{--G}^{OH} & & {}_{P}\text{AATTC--} \\ & 3'\text{--CTTAAG--}5' & \longrightarrow & \text{--CTTAA}_{P} & + & {}_{HO}\text{G--} \\ \textit{Pst}\text{I} & 5'\text{--CTGCAG--}3' & & \text{--CTGCA}^{OH} & & \text{pG--} \\ & 3'\text{--GACGTC--}5' & \longrightarrow & \text{--G}_{P} & + & {}_{HO}\text{ACGTC--} \end{array}$$

The protrusions generated by cleavage with one such enzyme are self-complementary and can associate spontaneously by base-pairing under appropriate conditions ("cohesive end" cleavage). In contrast, *Pvu*II cuts both strands at the same position (blunt or flush cleavage).

$$\begin{array}{llcllr} \textit{Pvu}\text{II} & 5'\text{--CAGCTG--}3' & & \text{--CAG}^{OH} & & \text{pCTG--} \\ & 3'\text{--GTCGAC--}5' & \longrightarrow & \text{--GTC}_{P} & + & {}_{HO}\text{GAC--} \end{array}$$

The distinction between cohesive and blunt cleavage is of paramount importance to DNA engineering.

Enzymes generating non-identical termini: this subclass of restriction enzymes includes those which recognise and cleave at a group of related sequences rather than at a single defined sequence. One such enzyme is *Ava*I which recognises the sequence CYCGRG where Y represents pyrimidine (T or C) and R represents purine (A or G). Thus the sequences CCCGGG, CTCGGG, CCCGAG and CTCGAG are all cleaved by this enzyme. When an asymmetric site is cleaved by *Ava*I the two ends, although complementary, are not identical.

$$\begin{matrix} 5'\text{--CTCGGG--}3' \\ 3'\text{--GAGCCC--}5' \end{matrix} \xrightarrow{AvaI} \begin{matrix} \text{--C}^{OH} \\ \text{--GAGCC}_{P} \end{matrix} + \begin{matrix} {}_{P}\text{TCGGG} \\ {}_{HO}\text{C--} \end{matrix}$$

A further example, *Bgl*I, cleaves DNA at the sequence GCCNNNN/NGGC. In other cases, the recognition sequence is not symmetrical and cleavage occurs at some distance to one side of the site recognised. For example, *Hga*I cleaves as shown below:

```
                ↓
5'--GACGCNNNNN NNNNNNN--3'
3'--CTGCGNNNNNNNNNN NN--5'
                   ↑
```

In these cases the exact termini produced by cleavage differ between one site and the next, and this is of particular importance when DNA fragments are to be joined end-to-end (see later sections).

B DNA ligases

The basic strategies of DNA manipulation involve enzymatic cleavage of DNA at defined sites followed by end-to-end joining of the various fragments produced. This second step is normally accomplished *in vitro* through the action of a DNA ligase (for a review see Engler and Richardson, 1982) though *in vivo* ligation after transformation occurs at a detectable frequency (Chang and Cohen, 1977).

Ligation consists of the joining of an extremity carrying a 5′-phosphate group to an extremity possessing a 3′-hydroxyl, thus reconstituting a normal internucleotide phosphodiester bond. Although DNA ligase is ubiquitous in living cells, only two enzymes are commonly used to carry out this reaction *in vitro*.

T4 DNA ligase. This is the enzyme that has been most widely used for the joining of DNA fragments. The relevant gene from bacteriophage T4 has been cloned in *E. coli* (Wilson and Murray, 1979) and the elevated production of ligase from this strain (Murray *et al.*, 1979) contributes to a reduction in the levels of contaminating activities in the purified enzyme. During the ligation reaction ATP is hydrolysed to AMP and PPi (Weiss and Richardson, 1968), while the phosphodiester bond generated derives from the 5′-phosphoryl group of the DNA substrate.

Bearing in mind the requirement for a 5′-phosphate and 3′-hydroxyl, the preferred substrate for T4 DNA ligase is a nicked ds DNA molecule. The enzyme is thus able to join cohesive termini generated by restriction enzyme cleavage, for base-pairing of the ss protrusions generates a substrate with staggered nicks in the two strands. The low temperature optimum of this joining reaction (4—15 °C, Ferretti and Sgaramella, 1981a) presumably reflects in part the stabilisation of the base-paired form.

The T4 enzyme, quite unlike other DNA ligases described in the literature, has the ability to link DNA molecules possessing "blunt" termini (Sgaramella *et al.*, 1970), although this "flush-end" reaction is some two orders of magnitude less efficient than "cohesive-end" joining (Sugino *et al.*, 1977). In addition, the T4 enzyme can carry out additional reactions such as the sealing of gaps in duplex DNA (Nilsson and Magnusson, 1982) and this low substrate specificity can lead to the generation of unexpected recombinant molecules.

It has been suggested that T4 RNA ligase is involved in the blunt-end joining activity associated with the DNA ligase (Sugino *et al.*, 1977), although the DNA ligase gene cloned in *E. coli* was subsequently shown to specify an enzyme with blunt-end joining activity (Murray *et al.*, 1979), demonstrating that this activity is an intrinsic property of the enzyme.

E. coli DNA ligase. E. coli ligase catalyses the repair of ss nicks in ds DNA but cannot catalyse end-to-end joining of DNA molecules with flush termini. The activity of the *E. coli* ligase, in contrast to the T4 enzyme, is substantially stimulated by the presence of NH_4^+ (Modrich and Lehman, 1973). In addition, the ligation reaction occurs concomitantly with the cleavage of NAD^+ to AMP+NMN rather than consuming ATP (Olivera and Lehman, 1967). The higher substrate specificity of the *E. coli* enzyme is useful in circumstances where nicks are to be sealed but where end to end joining of different molecules is to be avoided (Anderson, 1981). However, reports that the blunt-end joining activity of the T4 enzyme can be inhibited without reducing the cohesive-end joining activity (Ferretti and Sgaramella, 1981b) may lead to the increased use of the T4 enzyme in such circumstances. Strains over-producing the *E. coli* ligase have also been constructed (Panasenko *et al.*, 1978).

III Joining of fragments generated by restriction enzyme cleavage

A Basic DNA manipulations

The most central techniques of DNA manipulation are site-specific cleavage of a DNA molecule and subsequent rejoining of different DNA fragments to generate a recombinant DNA molecule. Such manipulations generally involve the specific cleavage of a DNA preparation (the "donor") with a restriction enzyme, mixing with a correspondingly cleaved vector molecule capable of autonomous replication, subsequent random end-to-end ligation *in vitro* followed by the transfer (transformation or transfection) of circular recombinant molecules into a suitable host. It must be stressed that, in many vector systems, circular molecules are necessary prerequisites for replication, and linear molecules are not normally recovered as clones.

The element of randomness during ligation reactions is particularly noteworthy, for much of the work involved during genetic manipulation lies in determining which of the "clones" produced contains the desired DNA fragments inserted in the correct position and orientation. There are, luckily, a number of strategies for ensuring that the correct hybrid molecules will predominate.

1 DNA concentration

In most ligation reactions the joining of extremity A to extremity B will be in competition with A-A and B-B fusions. Equally, in a 2 component mixture of donor (D) and vector (V), the proportion and yield of viable circular V + D molecules will depend on the absolute concentrations of both elements. Much mathematical analysis revolves about the apparent "j" value (Dugaiczyk *et al.*, 1975) of a molecule, which reflects the concentration of one end of a molecule in the immediate vicinity of the other. The shorter a DNA molecule, the more likely is it to circularise at any given concentration, and the fewer molecules which therefore participate in intermolecular ligation. To offset this problem, the concentration of the second DNA molecule is usually increased to compete this process, with the proviso that the probability of formation of linear oligomers is also increased. A more sophisticated strategy involves the ligation of only a fraction of the ends in the population, followed by dilution and religation to allow circularisation of joint molecules. Note that very small DNA fragments ($\leqslant$ 250 bp) are invariably cloned with high efficiency for their recircularisation is impaired on steric grounds (Shore *et al.*, 1981).

2 *"Sabotage" strategies*

The cloning of DNA fragments from a biological source *de novo* is performed relatively infrequently, and routine work more often consists of the transferring of a DNA fragment, defined by flanking restriction sites, from one location to another. Parental DNA molecules can often be eliminated by "sabotage" strategies. The simplest strategy involves the identification of a restriction recognition sequence in a parental or otherwise unwanted molecule but not in the molecule required. For instance, the fusion of restriction termini may eliminate the recognition sequences for both enzymes involved. Cleavage of the products of ligation with an appropriate enzyme prior to transformation will enrich the desired recombinant.

A more general procedure is to precleave the parental molecule containing the inserted target fragment with an enzyme lacking a recognition site within the target. Dephosphorylation (see below) of these termini and excision of the target fragment prior to religation very effectively eliminates the parental form.

3 *Cloning between dissimilar sites*

In cases where a fragment to be inserted into a circular vector molecule is flanked by dissimilar restriction termini, self-circularisation of the fragment is prohibited. Equally, the vector molecule can be cleaved by 2 different restriction enzymes at adjacent sites and the central fragment removed by physical techniques, thus ensuring incorporation of the target fragment. Kurtz and Nicodemus (1981) used this technique in conjunction with plasmid pBR322 to enhance the generation of recombinants having acquired an exogenous *Eco*RI-*Sal*I fragment; greater than 95% of the viable molecules recovered had incorporated a target DNA fragment.

4 *Dephosphorylation*

A vector DNA molecule cleaved at a single site with a restriction enzyme will normally be efficiently recircularised by DNA ligase. Thus, hybrid circular molecules having incorporated an exogenous DNA fragment will be in a minority. One successful solution has been to dephosphorylate the vector with either bacterial alkaline phosphatase or calf intestinal phosphatase. Since DNA joining mediated by DNA ligase requires a 5′ phosphoryl group, abortive recircularisation is prevented (Ullrich *et al.*, 1977); when both 5′ termini have been dephosphorylated, ligation can proceed in neither strand. However, at a junction between a donor and a vector molecule the donor is able to provide a single phosphoryl group suitable for ligation in one strand alone, and the residual nick is usually repaired after transformation by enzymes present *in vivo*. In certain cases it has been found advantageous to dephosphorylate the target DNA rather than the vector DNA (Ish-Horowicz and Burke, 1980) in order to prevent the cloning of more than one insert fragment per vector.

One note of caution: phosphatases are difficult to remove effectively and it is helpful to chelate their Zn^{2+} cofactor with EGTA and denature with phenol prior to subsequent manipulation.

B Homologous joining of restriction termini

We have discussed how the efficiency of end-to-end joining of DNA fragments is stimulated markedly by the formation of a substrate with staggered nicks in the two DNA strands. If the cohesive ends generated by cleavage do not match, then ligation is severely impaired. A fundamental technique of genetic manipulation takes advantage of the fact that certain enzymes generate identical cohesive ends even though they recognise different base sequences in DNA (Table 1). Although fusion of an *Eco*RI terminus with an identical terminus always regenerates the recognition sequence for *Eco*RI, recognition sequences are seldom regenerated when joining termini generated by dissimilar restriction enzymes.

An extension to the above technique is suggested by the finding that certain enzymes cleave at a distance from their recognition sequence. For example *Bbv*I cleaves as below:

```
5′-GCAGCNNNNNNNN↓
3′-CGTCGNNNNNNNNNNNN↑
```

Depending on the DNA sequence at the cleavage site, then cohesive ends may be generated compatible with the termini produced by a number of other enzymes. However, after site fusion by any method the recognition site remains intact (but not at both sides of the junction).

C Ordered ligation

Several restriction enzymes cleave ds DNA to give termini which are nonidentical (Section II.A). In such cases, ligase treatment of the

Table 1 Restriction enzymes generating identical cohesive ends. Restriction recognition sequences are given 5′→3′; the exact cleavage site being indicated by /. R and Y indicate purine (A or G) and pyrimidine (C or T) nucleotides respectively. In cases marked (*) the enzyme cleaves additionally at other sites to generate cohesive ends of different sequence.

Group	Enzyme	Cleavage sequence
I	*Sau* 3A	/GATC
	Bam HI	G/GATCC
	Bcl I	T/GATCA
	Bgl II	A/GATCT
	Xho II	R/GATCY
II	*Bss* HII	G/CGCGC
	Mlu I	A/CGCGT
III	*Taq* I	T/CGA
	Hpa II	C/CGG
	Sci NI	G/CGC
	Acc I	GT/CGAC (*)
	Acy I	GR/CGYC
	Asu II	TT/CGAA
	Cla I	AT/CGAT
	Nar I	GG/CGCC
IV	*Sal* I	G/TCGAC
	Xho I	C/TCGAG
V	*Nsp* I	RCATG/Y
	Sph I	GCATG/C
VI	*Hgi* AI	GTGCA/C (*)
	Pst I	CTGCA/G
VII	*Bde* I	GGCGC/C
	Hae II	RGCGC/Y
VIII	*Cfr* I	Y/GGCCR
	Xma III	C/GGCCG

digest may often result in ordered reassembly of the original DNA. For example, Hartley and Gregori (1981) took advantage of an asymmetric *Ava* I site to generate an ordered polymer of a restriction fragment. Also, Doel *et al.* (1980) polymerised a short duplex segment with non-identical cohesive ends to generate a repetitive gene coding for a polymeric form of the sweetener, L-aspartyl-L-phenylalanine.

Cloning between dissimilar sites (Section III.A) and ordered ligation use the same principle to eliminate classes of possible recombinants. This strategy can also be used when cloning from partial restriction enzyme digests. Here it is advantageous to cleave completely at an outside site A, followed by extremely mild digestion with the second enzyme (B) to introduce about one cut per molecule. Cloning between vector sites A and B gives an ordered array of partial digest recombinants.

A technical note: the amount of enzyme required to cut a proportion of sites is disproportionately reduced over that necessary for cutting all the sites, and 10—20 fold less enzyme is often appropriate.

D Non-homologous joining of restriction termini

It is often necessary to join DNA fragments with non-complementary termini. In such cases ligation is extremely inefficient, although partial complementarity can sometimes allow successful ligation. For instance, an assymetric *Ava*I terminus may be ligated, at low efficiency, to a terminus generated by *Sal*I cleavage (see below), the resulting mismatch

$$\begin{array}{l} --G^{OH} \\ --CAGCT_{P} \end{array} + \begin{array}{r} {}_{P}TCGGG-- \\ {}_{HO}C-- \end{array} \longrightarrow \begin{array}{l} --GTCGGG-- \\ --CAGG_{T}C-- \end{array}$$

being corrected *in vivo* (mismatch cloning). In another case, the ss cohesive end at one terminus may base-pair with part of the cohesive end at a dissimilar terminus (overlap cloning). For example, termini generated by enzymes of group II (Table 1) may be joined to termini generated by enzymes of group III, and the cohesive end produced by e.g. *Mlu*I will ligate to that produced by e.g. *Cla*I.

$$\begin{array}{l} --AT^{OH} \\ --TAGC_{P} \end{array} + \begin{array}{r} {}_{P}CGCGT-- \\ {}_{HO}A-- \end{array} \longrightarrow \begin{array}{l} --ATCGCGT-- \\ --TAGC_{P\,HO}A-- \end{array}$$

Here the gap left in one strand is filled-in *in vivo*. Equally, termini generated by *Pvu*I (recognition sequence CGAT/CG) may ligate to *Eco*RV termini (GATAT/C), and *Sac*II termini (CCGC/GG) to termini generated by enzymes of group VII (GCGC/).

In most cases, however, the ss protrusions may hinder rather than aid the ligation reaction. One commonly used strategy is to first remove the ss protrusions and subsequently employ the "blunt-end" activity of T4 DNA ligase to join the fragments. This can be accomplished in two ways, either with ss-specific nucleases or one of a number of DNA polymerases as discussed below.

1 Trimming with single-strand specific nucleases

Three nucleases are commonly used to remove ss extensions from a DNA terminus. Mung-bean endonuclease is not completely specific in its preference for single-stranded DNA (see Laskowski, 1980) and has been least used for this purpose. *Neurospora crassa* nuclease is not naturally specific for ss DNA and the enzyme requires pretreatment to remove double-stranded DNase activity (Fraser, 1980). The enzyme of choice is nuclease S1 which can be purified to homogeneity (Vogt, 1973). Under suitable ionic conditions ($>$ 200 mM NaCl) this enzyme is extremely specific for single-stranded DNA and has an optimal activity at close to pH 4 (Vogt, 1980) though a higher pH is often employed to reduce problems of depurination. Its requirement for a Zn^{2+} cofactor, rather than Mg^{2+}, allows digestion to be performed in the absence of Mg^{2+}, thus rendering contaminant ds nucleases inactive. A note of caution — S1 nuclease at high concentration will cleave ss regions of partial denaturation in duplex DNA (Beard *et al.*, 1973; Lilley, 1981) and for this reason digestion with S1 is usefully performed at a lower temperature (25°C) to avoid digestion at transiently denatured ends of a DNA fragment. Unfortunately the DNA termini produced after S1 digestion are rarely precisely flush and subsequent repair (see below) is advantageous (Shishido and Ando, 1981).

2 *Repair synthesis with DNA polymerase*

Three polymerases have routinely been employed to carry out this reaction. *E. coli* DNA polymerase I has been extensively characterised (for a review see Lehman, 1981a) and a proteolytic fragment of this polymerase, the "Klenow fragment" (Klenow and Henningsen, 1970; Setlow *et al.*, 1972) is widely available. A second enzyme, isolated from T4-infected *E. coli*, T4 DNA polymerase, has also been much used (see Lehman, 1981b). The reverse transcriptase isolated from virions of avian myeloblastosis virus (see Verma, 1981) has been used by some workers. These enzymes are compared in Table 2. From the point of view of DNA engineering, two reactions are of interest.

Table 2 Comparison of different DNA polymerases useful in modifying DNA.

Enzyme	5′→3′ polymerisation	3′→5′ exonuclease	5′→3′ exonuclease
E. coli DNA polymerase I	+	+	+
Polymerase I 'Klenow fragment'	+	+	—
T4 DNA polymerase	+	+	—
AMV reverse transcriptase	+	—	—

(1) End filling. If a terminus generated by restriction enzyme digestion possesses a 5′ protruding ss extension (for example, the termini generated by *Eco*RI, *Hin*dIII or *Bam*HI digestion) then incubation with any one of the enzymes described above in the presence of the 4 dNTPs will "fill-in" the extension. This method has been widely used to render such ends flush.

$$\begin{array}{lcl} --\mathrm{G}^{\mathrm{OH}} & \xrightarrow[\text{DNA polymerase}]{\text{4dNTP}} & --\mathrm{GGATC}^{\mathrm{OH}} \\ --\mathrm{CCTAG_P} & & --\mathrm{CCTAG_P} \end{array}$$

(2) Trimming back. The *E. coli* and T4 enzymes also possess a very useful 3′→5′ exonucleolytic activity which can be used to remove the 3′ protruding ss extensions generated by cleavage with certain enzymes (e.g. *Pst*I, *Kpn*I). Prolonged incubation in the presence of the 4 dNTPs selectively removes the ss extension, while further digestion is effectively arrested by the synthesis which occurs preferentially in the presence of the complementary strand and the dNTPs.

$$\begin{array}{lcl} --\mathrm{CTGCA}^{\mathrm{OH}} & \xrightarrow[\substack{\textit{E. coli}\text{ polymerase} \\ \text{T4 polymerase}}]{\text{4dNTP}} & --\mathrm{C}^{\mathrm{OH}} \\ --\mathrm{G_P} & & --\mathrm{G_P} \end{array}$$

Incubation in the absence of triphosphates leads to continued exonucleolytic digestion at the 3′ end which can be halted at specific sites by the addition of one or more triphosphates (see for instance Ciampi *et al.*, 1982).

The combination of the filling and trimming reactions has been used widely to convert "ragged" ends to blunt ends using either *E. coli* DNA polymerase (Mulligan *et al.*, 1979; Anderson, 1981) or the T4 enzyme (Wartell and Reznikoff, 1980).

E Reconstructing restriction sites

Using DNA polymerase and S1 nuclease, separately or in combination, much can be done to manipulate the DNA sequence at a junction generated during the end-to-end joining of two DNA fragments. By using selected deoxynucleoside triphosphates, particular ends can be filled-in to different extents (see for instance Donoghue and Hunter, 1982). One simple application of this technique is to treat restriction termini with DNA polymerase in the presence of one or more dNTPs and to religate that terminus to itself to generate a site of different specificity (Table 3). Equally, after filling in a restriction site with a single deoxynucleoside triphosphate, subsequent S1 digestion and self-ligation can be used to change the specificity of the junction. Shepard *et al.* (1982) repaired an *Xba*I

Table 3 Conversion of restriction sites by filling-in and resealing. In each case the central 4 bp sequence of the recognition site is given; the position of cleavage is not indicated since the exact position varies depending on which enzyme is used. In each case the restriction site is filled-in by treatment with DNA polymerase in the presence of one or more dNTPs, and religated, to itself, to generate a different restriction recognition sequence. For instance, the site Y/GGCCR can be cleaved with *Cfr*I, filled-in with DNA polymerase, dGTP and dCTP and religated to itself to generate YGGCCGGCCR. This includes the recognition sequence for *Nae*I (GCCGGC). In a second example, the sequence TCGA may be either cleaved by *Taq*I (T/CGA), filled-in and religated to generate a *Nru*I site (TCGCGA), or alternatively, if it forms part of the recognition sequence of a *Sal*I site (G/TCGAC), cleavage with this enzyme, filling-in and religation will generate GTCGATCGAC which includes the recognition sequence for *Pvu*I (CGATCG). Enzymes in brackets additionally cleave other recognition sites. (*) indicates that *E. coli dam* methylation (see section V) blocks digestion of the sequence given by the enzyme specified.

4 sequence	--->	6 sequence	--->	8 sequence
GGCC *Cfr*I *Xma*III				GGCCGGCC *Nae*I
AGCT *Hin*dIII		AGCGCT *Hha*I *Hae*II		
GCGC *Sci*NI *Nar*I (*Acy*I) (*Hgi*CI) *Mlu*I *Bss*HII		GCGCGC *Tha*I *Bss*HII		GCGCGCGC *Bss*HII
ACGT (*Acy*I)		ACGCGT *Tha*I *Mlu*I		
TCGA *Taq*I *Asu*II *Cla*I *Sal*I *Xho*I (*Acc*I) (*Ava*I)		TCGCGA *Tha*I *Nru*I		TCGATCGA *Sau*3A *Pvu*I

Table 3 (*Continued*)

4 sequence	---→	6 sequence	---→	8 sequence
CCGG *Hpa* II *Xma* I (*Ava* I)		CCGCGG *Tha* I *Sac* II		CCGGCCGG *Hae* III *Xma* III
GATC *Sau* 3A *Bam* HI *Bcl* I (*) *Bgl* II *Xho* II		GATATC *Eco* RV		GATCGATC *Sau* 3A *Taq* I (*) *Cla* I (*)
GTAC (*Hgi* CI)		GTATAC *Sna* I *Acc* I		
CTAG *Xba* I				CTAGCTAG *Alu* I

terminus in the presence of dCTP alone. S1 treatment and religation led to the generation of a new site for *Taq* I.

```
-TCTAGA-     XbaI       -T OH          pCTAGA-
-AGATCT-  ---------->   -AGATCp     +      HOT-

      dCTP              -TC OH           PGA-
 ------------->                     +
Klenow polymerase       -AGp           HOCT-
S1 nuclease

   DNA ligase           -TCGA-
 ------------->         -AGCT-
                         ****
                                Taq I recognition sequence
```

An extension of this technique is to ligate a filled-in restriction terminus to one which has been treated with S1 nuclease to remove its single-stranded extension. For example, an *Eco* RI terminus (cleavage site G/AATTC) treated with DNA polymerase can be ligated to a *Bam* HI terminus (cleavage site G/GATCC) which has been digested with S1 to regenerate the recognition sequence for *Eco* RI.

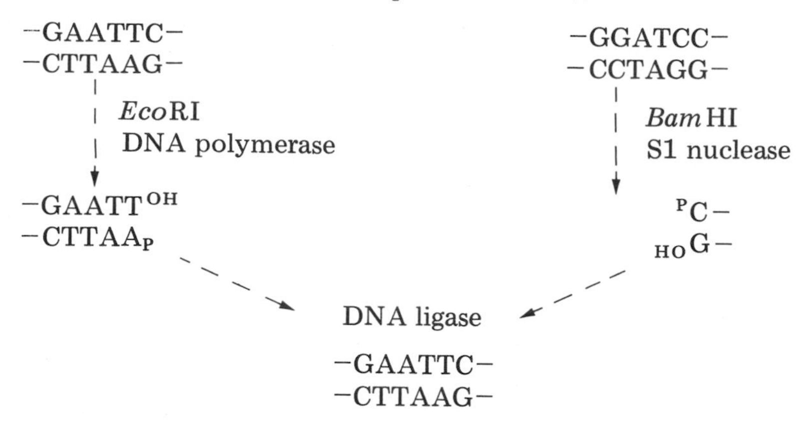

Reversing the treatment leads to the regeneration of a *Bam*HI site.

Further, extremities which have been filled-in by DNA polymerase treatment can be ligated to other termini generated by flush-end cleavage to regenerate the restriction recognition sequence. In this way a filled-in cohesive *Eco*RI terminus (–GAATT$^{3'}$) ligated to a flush *Pvu*II terminus ($^{5'}$CTG––) regenerates an *Eco*RI recognition sequence (GAATTCTG). In the same way, two dissimilar filled-in cohesive termini (for instance *Eco*RI, –GAATT$^{3'}$, and *Xba*I, $^{5'}$CTAGA–) can be fused to regenerate both recognition sequences (–*GAATTC*TAGA–).

The reconstruction of a restriction recognition site is often the only easy means for determining whether a given construction has been successful. Thus, all combinations of filling and trimming should be carefully examined for possible regeneration of recognition sequences (see Table 4). By reconstructing restriction sites, the combination of the site manipulations discussed here with the synthetic

---

Table 4 Reconstruction of restriction recognition sequences at junctions of restriction termini. A lists restriction enzymes generating 5′ss cohesive ends suitable for filling-in and site reconstruction after ligation. In all cases cited, filled-in ends lack a single terminal nucleotide from the recognition sequence. In B the various defined means are listed for providing the missing oligonucleotide. In each case the cleavage is either blunt (B) or cohesive where subsequent filling-in (F) or trimming (T) are required to give blunt ends.

Combinations of filled-in termini from A with suitably-treated termini from B (in the same column) regenerate the recognition sequence for the enzyme in A. Pairs of enzymes in A and B marked * reconstruct both recognition sequences. In the 3 cases marked 1 self-ligation after filling-in results in tandem duplication of the recognition sequence. Only enzymes with hexameric recognition sequences are listed in the Table, and only appear when reconstruction is effective at all sites. Similar site reconstructions are possible using restriction enzymes with tetrameric, redundant pentameric or hexameric recognition sequences (e.g. *Taq*I, *Hinf*I and *Ava*I respectively).

A. Restriction recognition sequences reconstructed by cleavage, filling-in and ligation to the 4 possible 5′ terminal nucleotides

G	A	T	C
* *Afl* II	*Bcl* I	* *Bgl* II	* *Bam* HI
*** *Ava* I	*Cfr* I	*Hin* dIII	1 *Bss* HII
1 **** *Cfr* I	* *Xba* I	** *Mlu* I	* *Bst* EII
** *Nco* I		* *Xho* II	** *Eco* RI
* *Xho* I			* *Hgi* CI
** *Xma* I			*** *Sal* I
1 *Xma* III			* *Xho* II

B. Enzymes generating each 5′ terminal base (below) after filling-in (F), trimming (T) or blunt-end cleavage (B)

G	A	T	C
T *Afl* II	T *Acc* I	* F *Afl* II	F *Acy* I
T *Ava* I	B *Aha* III	T *Bgl* II	T *Apa* I
** F *Bam* HI	T *Asu* II	T *Hin* dIII	F *Asu* II
F *Bcl* I	T *Bcl* I	T *Mlu* I	B *Bal* I
* F *Bgl* II	T *Cla* I	F *Mst* II	T *Bam* HI
** F *Bst* EII	* F *Eco* RI	T/F *Nde* I	1T/F *Bss* HII
1 F *Cfr* I	F *Hin* dIII	** F *Sal* I	T *Bst* EII
** F *Hgi* CI	B *Hpa* I	* F *Xho* I	F *Cla* I
B *Mst* I	T *Xba* I		T *Eco* RI
T *Mst* II			T *Eco* RV
B *Nae* I			B *Hae* I
T *Nco* I			T *Hgi* AI
T *Pst* I			T *Hgi* CI
T *Sac* II			T *Hgi* JII
B *Sma* I			T *Kpn* I
T *Xho* I			*** F *Mlu* I
*** F *Xho* II			T/F *Nar* I
T *Xma* I			* F *Nco* I
**** 1 T/F *Xma* III			T *Pvu* I
			B *Pvu* II
			T *Sac* I
			T *Sal* I
			T *Sph* I
			B *Stu* I
			** F *Xba* I
			* F *Xma* I

oligonucleotide technology described in the next section provides an extremely potent method for monitoring events taking place at the molecular level.

A sophisticated example of this technique deserves mention. Panayotatos and Truong (1981) used a double-strand exonuclease to trim back from a restriction site and were able to recognise specific deletions by the reconstruction of a particular recognition site after religation to a filled-in restriction terminus. Such methods are described in greater detail in section VIII.A.

IV Oligonucleotide linkers and adaptors

Due to improvements in synthetic techniques, artificial oligonucleotides have met increasing use in DNA engineering. In this section we restrict discussion to oligonucleotides useful for linking DNA fragments.

A Linkers

Linkers are synthetic ss oligonucleotides which self-associate to form symmetrical ds molecules containing the recognition sequence for a restriction enzyme (Greene *et al.*, 1975; Bahl *et al.*, 1976). For example, the oligonucleotide dCCGAATTCGG will self-associate to give a duplex structure containing the *Eco*RI recognition sequence (indicated by asterisks).

```
       * *****
5′ CCGAATTCGG 3′
3′ GGCTTAAGCC 5′
```

These oligonucleotides can be used in two distinct manners depending whether the 5′ terminus of the linker is phosphorylated or non-phosphorylated.

1 Phosphorylated linkers

In the classic approach to using linkers, the unphosphorylated oligonucleotide deriving from chemical synthesis is 5′-phosphorylated using polynucleotide kinase and ATP. Such linkers can then be ligated to the flush end of a DNA fragment using T4 DNA ligase. In fact, under standard conditions an excess of linker is employed to drive the reaction and a linear polymer of ds linkers is formed at both ends of the DNA fragment. Complete digestion of the ligation products with the cognate restriction enzyme generates a DNA fragment flanked with cohesive ends (Bahl *et al.*, 1976; Heynecker *et al.*, 1976) suitable for direct insertion into a similarly cleaved vector molecule. One disadvantage of the method, namely that linkers cannot be used in conjunction with a DNA fragment possessing an internal recognition sequence for the enzyme in question, was overcome by Maniatis *et al.* (1978) who used a specific DNA methylase *in vitro* to protect internal sites from cleavage (see later sections).

2 *Non-phosphorylated linkers*

An alternative strategy is to use linkers lacking 5′ terminal phosphate groups. Although the method may be suitable for recombining different DNA molecules, non-phosphorylated linkers have been most useful in cases where a new restriction site is to be inserted into a circular DNA molecule (Lathe, R. unpublished). Ligation of a blunt-ended DNA fragment with unphosphorylated linkers leads to the formation of a covalent bond, in one strand alone, between each terminus and the linker (Fig. 1). Removal of the unligated linker strand and rehybridisation of the now cohesive ends leads to the

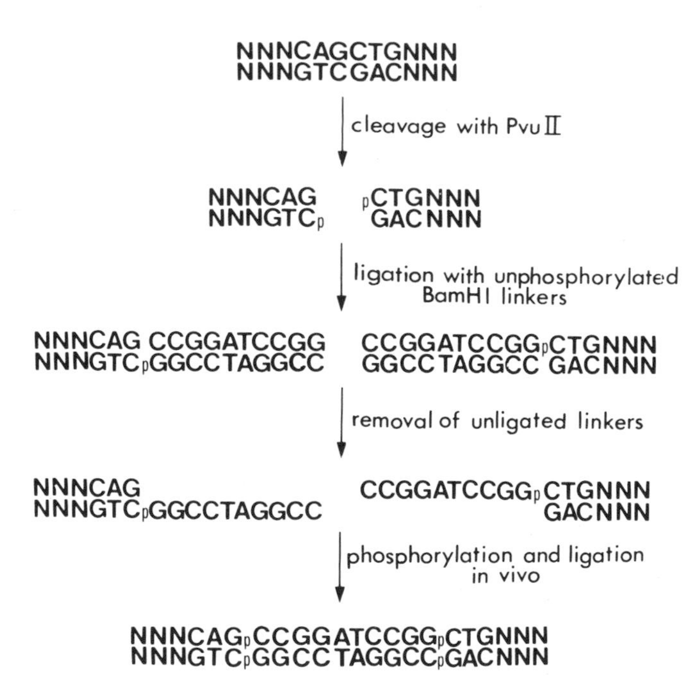

Figure 1 Insertion of a non-phosphorylated *Bam*HI into the *Pvu*II site of plasmid pBR322. Blunt-end ligation of the synthetic duplex linker to termini generated by *Pvu*II cleavage involves the formation of covalent links in only one strand at each junction. Removal of unligated oligonucleotides generates a 10 bp overlap allowing *in vivo* circularisation and religation of the plasmid.

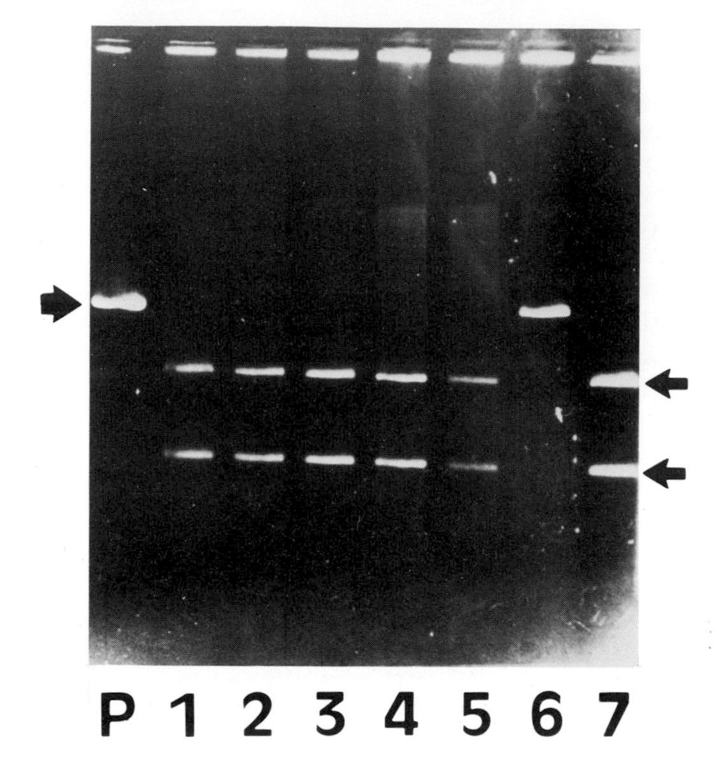

Figure 2 Products of non-phosphorylated *Bam* HI linker insertion into the *Puv* II site of pBR322. The plasmid was cleaved with *Pvu* II, ligated with an excess of unphosphorylated decameric *Bam* HI linkers, heated briefly to 70°C and precipitated with spermine to remove unligated linkers. Plasmids isolated from ampicillin resistant colonies obtained after transformation of this material into *E. coli* were digested with *Bam* HI and electrophoresed on an agarose gel (migration above to below). P contains pBR322 DNA cleaved once with *Bam* HI; slots 1–7 contain *Bam* HI-treated plasmid DNAs obtained from independent transformants. In 6/7 cases a second *Bam* HI cleavage site has been introduced, giving rise to 2 bands on the gel.

formation of a circular molecule with staggered nicks which can be repaired *in vivo*. This method is particularly useful where the molecule contains an internal recognition sequence for the enzyme in question. For instance, the method has been used to replace the *Pvu* II site in plasmid vector pBR322 (Bolivar *et al.*, 1977) by a *Bam* HI site, generating a plasmid containing 2 *Bam* HI sites (Fig. 2).

B Half-linkers

Half-linkers are formally equivalent to linker oligonucleotides which have been precleaved by the cognate restriction enzyme. They consist of a ds DNA molecule comprising one flush end which can be ligated to a blunt-ended molecule, and one cohesive end suitable for direct cloning into a restriction site (Bahl *et al.*, 1978; Wu *et al.*, 1978). Restriction enzyme cleavage after ligation is avoided but the method normally requires that one terminus be non-phosphorylated to prevent abortive self-ligation of the preformed cohesive ends. Norris *et al.* (1979) successfully used this method to clone a fragment of MS2 DNA into a plasmid vector.

C Adaptors

Decameric adaptors: Bahl *et al.* (1978) described how a decameric single-stranded oligonucleotide could be used to fuse a terminus carrying a 3′-protruding cohesive end to a 5′-protruding cohesive end.

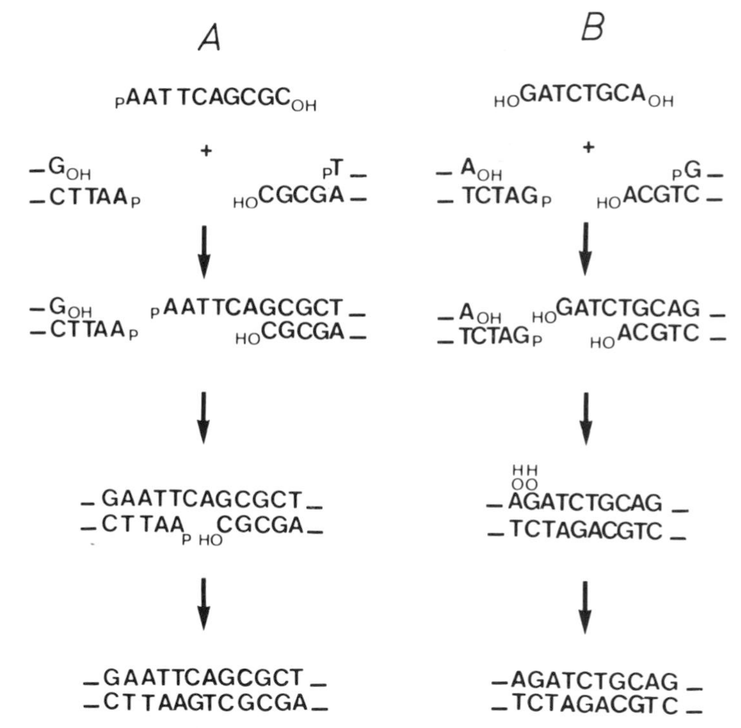

Figure 3 Fusion of restriction sites using ss adaptor oligonucleotides. In *A*, termini generated by *Eco*RI and *Hae*II cleavage are fused in two ligation steps to generate a junction containing a 2 bp gap in one strand; repair *in vivo* leads to the regeneration of both the *Eco*RI (GAATTC) and *Hae*II (AGCGCT) recognition sequences (Bahl *et al.*, 1978). In *B*, termini generated by *Bgl*II and *Pst*I cleavage are similarly linked in two steps to generate a hybrid terminus containing a dephosphorylated nick in one strand. Phosphorylation and ligation *in vivo* regenerates overlapping *Bgl*II (AGATCT) and *Pst*I (CTGCAG) recognition sequences (Lathe *et al.*, 1982).

In their strategy, the ss oligonucleotide ligates to the dissimilar extremities of the two DNA fragments to reconstruct both recognition sequences (Fig. 3A), the gap left in one strand being repaired *in vivo*.

Octameric adaptors: Lathe *et al.* (1982) describe an extension to the method of Bahl *et al.* (1978) where octameric adaptor oligonucleotides are used to fuse dissimilar termini, again with the proviso that the termini have 5′ and 3′ protruding cohesive ends (Fig. 3B). One advantage presented by the use of octameric (and equivalent shorter oligonucleotides) is that initial ligation of the oligonucleotide consumes the terminal phosphate and prevents subsequent ligation of identical termini. In the case of decameric adaptors the same effect is accomplished by the residual gap in one strand. Table 5 summarises the restriction site reconstructions permitted using ss oligonucleotides of the second type.

D Connectors

A final method for joining defined termini involves the use of 2 adaptors in combination. In the scheme proposed by Bahl *et al.* (1978), two adaptors possessing a complementary core region are added to two different restriction termini possessing ss extensions in the same sense (see below):

```
−G^OH           pAATTCCCGGG_OH         pTCGAC−
−CTTAA_P    +                       +     HO G−
EcoRI          HO GGGCCCAGCT_P        SalI
terminus              |               terminus
                      |
                      ↓ DNA ligase
                  SmaI site
                   ******
            −GAATTCCCGGGTCGAC−
            −CTTAAGGGCCCAGCTG−
             ******     ******
            EcoRI site   SalI site
```

An advantage of this technique is that an additional restriction recognition sequence may be introduced between the two sites (see Rothstein *et al.*, 1979).

E Other uses of linkers

Linkers and other small artificial duplex DNAs have found an extraordinarily wide range of applications. Charnay *et al.* (1978) used

Table 5 Reconstruction of restriction recognition sequences after terminal fusion using a single-stranded oligonucleotide adaptor. The restriction enzymes listed on the left generate termini with a 5′ ss extension, those above a 3′ ss extension; in each case the recognition sequence and exact cleavage sites are presented. R = purine nucleotide; Y = pyrimidine nucleotide. Site fusions performed with appropriate single-stranded adaptors may lead to the recreation of the recognition sequence for the enzyme generating the 5′ ss extension (≪) or the sequence for the enzyme generating the 3′ extension (≫). Cases where both restriction recognition sequences are regenerated are marked *. An example of such a fusion is presented in Fig. 3B.

	*Eco*RV GATAT/C	*Hae*II RGCGC/Y	*Kpn*I GGTAC/C	*Pst*I CTGCA/G	*Pvu*I CGAT/CG	*Sac*I GAGCT/C	*Sph*I GCATG/C
*Ava*I							
C/CCGGG	≫	*	*			≫	≫
C/TCGGG	≫	*	*			≫	≫
C/CCGAG		*	≪				
C/TCGAG		*	≪				
*Bam*HI							
G/GATCC				≫			≪
*Bgl*II							
A/GATCT				*			
*Cla*I							
AT/CGAT	*	≫	≫		*	≫	
*Eco*RI							
G/AATTC							≪
*Hin*dIII							
A/AGCTT				≪			
*Sal*I							
G/TCGAC		≫					≪
*Xba*I							
T/CTAGA	*	≫	≫		≪	*	≫
*Xho*I							
C/TCGAG		*	≪				
*Xma*I							
C/CCGGG	≫	*	*			≫	≫

sequential S1 treatment and linker insertion to alter progressively the phase of translation at a unique *Eco*RI recognition site in the N-terminal region of the β-galactosidase gene inserted into a plasmid vector. After cleavage with *Eco*RI and S1 digestion, octameric linkers were ligated at the termini of the plasmid. Recircularisation of the plasmid after subsequent *Eco*RI cleavage gave an altered phase of translation at the *Eco*RI site, and exact repetition of the construction using this new vector as the starting point generated a third phase of translation. Thus plasmids were constructed suitable for the insertion of an unknown protein coding sequence, leading to in-phase translation in one of the three vectors.

Gray *et al.* (1982) used a short synthetic duplex with dissimilar cohesive termini at the two ends to fuse the coding sequence for human gamma interferon to an active *E. coli* promoter sequence. The nucleotide sequence between the *Eco*RI and *Bst*NI cohesive ends used in this study was designed so as to code for the missing N-terminal amino acids of the mature interferon polypeptide. Jay *et al.* (1981) used a similar duplex with non-identical cohesive ends to insert, upstream of the small t (tumour) antigen of simian virus 40, a synthetic translation initiation sequence for *E. coli* ribosomes.

F Technical note

Two problems are often encountered when handling oligonucleotide linkers and adaptors. The first arises because a molar excess of oligonucleotide is employed to ensure that all target DNA molecules ligate to a linker. An unfortunate consequence is that the excess oligonucleotide competes with the modified target molecule in subsequent steps. Various strategies have been employed to remove excess oligonucleotides, including gel filtration (Bahl *et al.*, 1978), gel electrophoresis (Norris *et al.*, 1979) and ultrogel chromatography (Anderson, 1981) though precipitation with spermine (Hoopes and McClure, 1981) is likely to supplant these methods.

The second problem is often encountered with long self-symmetrical linkers. These oligonucleotides can both form a duplex or alternatively a folded-back hairpin structure (below).

$$\begin{array}{lcl} {}^{\mathrm{P}}\mathrm{GCCGGATCCGGC}^{\mathrm{OH}} \longleftarrow & & \longleftarrow \; {}^{\mathrm{P}}\mathrm{GCCG}^{\mathrm{G}}\mathrm{A} \\ {}_{\mathrm{HO}}\mathrm{CGGCCTAGGCCG}_{\mathrm{P}} \longrightarrow & {}^{\mathrm{p}}\mathrm{GCCGGATCCGGC}_{\mathrm{OH}} & \longrightarrow \; {}_{\mathrm{HO}}\mathrm{CGGCC}_{\mathrm{C}}\mathrm{T} \end{array}$$

Care must therefore be taken to pre-anneal linker oligonucleotides prior to use to favour the linear duplex form. Further, this suggests that a ss linker could be devised to fold back upon itself to give a substrate for ligation and restriction enzyme cleavage. The use of such a molecule is likely to have a number of technical advantages.

V DNA methylation

The DNA of many organisms is methylated *in vivo*, by and large at the 5 position of cytosines or at the N6 position of adenines (for a review, see Hattman, 1981). Apart from the necessary protection against cleavage of self-DNA in organisms producing restriction enzymes, the exact role of DNA methylation is not understood. In many microorganisms, sequence-specific DNA methylation is an essential component of the restriction-modification system, but just as methylation *in vivo* protects DNA from digestion, so it may protect DNA *in vitro* from the attentions of the genetic engineer.

A Methylation in eukaryotes

The subject of DNA methylation in eukaryotes is extraordinarily complex, and only aspects of importance to genetic engineering will be discussed here. The DNA of, in particular, vertebrates and plants but not insects, is highly resistant to digestion by enzymes such as *Hpa*II and *Sma*I. It has now emerged that this phenomenon is due to two separate but linked phenomena.

1. The recognition sequences for both enzymes contain the dinucleotide pair 5′-CG-3′, and this sequence is often methylated at the 5 position on the cytosine in many organisms (see for instance Grippo *et al.*, 1968). Methylation of this sequence prevents cleavage by these enzymes.
2. The dinucleotide CG is selectively depleted in the DNA of higher organisms, probably due to the hyper-mutability of the methylated C (Bird, 1980). Therefore restriction recognition sequences containing CG are also selectively depleted.

To a certain extent problems of DNA methylation can be avoided by judicious choice of restriction enzymes (see Table 6), but we must stress that no aspect of DNA methylation in higher eukaryotes is fully understood, and one must expect that other restriction sites may be blocked even though they do not contain the CG pair. For example, it has been reported that rye DNA is partially resistant to cleavage with *Pst*I (Bedbrook and Gerlach, 1980). Although eukaryotes are not generally thought to possess restriction-modification systems, certain yeasts have recently been shown to possess site-specific endodeoxyribonucleases (Watabe *et al.*, 1981) and site-specific methylation is to be anticipated.

B Methylation in prokaryotes

Since the vast majority of genetic engineering involves the passage of DNA at some time or other into the bacterium *E. coli*, this section will concentrate on DNA methylation in this organism. Nevertheless, it must not be forgotten that all other common prokaryotic hosts for gene cloning do, in general, have their own restriction-modification systems (see Brooks and Roberts, 1982).

Escherichia coli K12, the strain most commonly used, contains at least 3 distinct methylation activities.

1 *The dam methylase*

Most K12 strains possess a DNA adenine methylase (*dam*) which adds a methyl group to the N6 position of adenine in the sequence 5′-GATC-3′ (Lacks and Greenberg, 1977; Geier and Modrich, 1979). Unlike many bacterial methylation enzymes, there seems to exist no corresponding restriction enzyme in *E. coli* which recognises this sequence. Instead, *dam* methylation is probably involved in strand selection during mismatch repair (Radman *et al.*, 1980). *Dam* methylation is responsible for the resistance of DNA extracted from *E. coli* to restriction endonuclease *Bcl*I (see Table 6) and for the inability of *Taq*I (recognition sequence TCGA) to cleave an overlapping *Taq*I:*dam* sequence TCG^{m}ATC in plasmid pBR322 (Backman, 1980). Similarly *Cla*I sites (recognition sequence ATCGAT) are very often encountered which are resistant to digestion — in *E. coli* K12 one site in four will be methylated at the second adenine such as to block restriction. The effect of methylation at the first adenine is unknown. Problems of overlapping restriction recognition sequences and *dam* methylation sites are discussed by Backman (1980) and Venegas *et al.* (1981). Fortunately, strains deficient in the *dam* methylase do exist (Marinus, 1973) and DNA can be passaged through such a strain in cases where a defined cleavage by such an enzyme must be made, although *dam* strains usually transform poorly.

2 *The dcm methylase*

Extracts of *E. coli* K12 contain a second methylase which methylates C residues at the 5 position. The *dcm* (DNA cytosine methylase) enzyme recognises the sequence CC(A/T)GG and methylates the second C (May and Hattman, 1975; Buryanov *et al.*, 1978). Restriction enzyme *Eco*RII recognises the same DNA sequence (Boyer *et al.*, 1973) and *dcm* methylation blocks its activity (Hattman, 1977). Other restriction recognition sequences which could potentially be rendered resistant by *dcm* methylation include those for *Bam*HI (GGATCC), *Bal*I (TGGCCA), *Kpn*I (GGTACC), *Nar*I (GGCGCC) and *Stu*I (AGGCCT). As before, strains deficient in the *dcm* methylase are available (Marinus, 1973).

Table 6 Prevention of restriction enzyme cleavage by site-specific methylation. Recognition sequences and methylation sites blocking restriction are given for a number of restriction enzymes. Data from Roberts (1981), McClelland (1981a), Brooks and Roberts (1982) and references therein. Sites methylated in *E. coli* K12 are marked *, sites only methylated in certain instances (*).

Restriction enzyme	Sequences cut	Sequences not cut
*Eco*RI	GAATTC	GA$\overset{m}{\dot{A}}$TTC
*Hin*dIII	AAGCTT	$\overset{m}{\dot{A}}$AGCTT
		A$\overset{m}{\dot{A}}$GCTT
*Bam*HI	GGATCC	GGAT$\overset{m}{\dot{C}}$C
	GG$\overset{m}{\dot{A}}$TCC *	
*Bgl*II	AGATCT	AGAT$\overset{m}{\dot{C}}$C
	AG$\overset{m}{\dot{A}}$TCT *	
*Pvu*I	CGATCG	CGAT$\overset{m}{\dot{C}}$G
	CG$\overset{m}{\dot{A}}$TCG *	
*Bcl*I	TGATCA	TG$\overset{m}{\dot{A}}$TCA *
*Sau*3A	GATC	GAT$\overset{m}{\dot{C}}$
	G$\overset{m}{\dot{A}}$TC *	
*Dpn*I	G$\overset{m}{\dot{A}}$TC *	GATC
*Mbo*I	GATC	G$\overset{m}{\dot{A}}$TC *
*Sma*I	CCCGGG	CC$\overset{m}{\dot{C}}$GGG
*Xma*I	CCCGGG	
	CC$\overset{m}{\dot{C}}$GGG	
*Hpa*II	CCGG	C$\overset{m}{\dot{C}}$GG
	$\overset{m}{\dot{C}}$CGG	
*Msp*I	CCGG	$\overset{m}{\dot{C}}$CGG
	C$\overset{m}{\dot{C}}$GG	
*Eco*RII	CC(A/T)GG	C$\overset{m}{\dot{C}}$(A/T)GG *
*Cla*I	ATCGAT	ATCG$\overset{m}{\dot{A}}$T (*)
*Sal*I	GTCGAC	GTCG$\overset{m}{\dot{A}}$C
*Xho*I	CTCGAG	CT$\overset{m}{\dot{C}}$GAG
		CTCG$\overset{m}{\dot{A}}$G
*Taq*I	TCGA	TCG$\overset{m}{\dot{A}}$ (*)

3 *The hsd methylase*

This methylase forms a component of the *E. coli* K12 type I restriction/modification system and is encoded by the *hsd*M gene.

The sequence recognised is most probably as below (Kan *et al.*, 1979) and adenine methylation at such sites (Haberman *et al.*, 1972) could potentially block the activity of enzymes with overlapping recognition sequences.

5′---AACNNNNNNGTGC---3′
3′---TTGNNNNNNCACG---5′

Burckhardt *et al.* (1981) describe a *Hpa*I (recognition sequence GTTAAC) site in bacteriophage lambda which is blocked by K-specific *hsd* methylation; nevertheless, many *E. coli* strains employed in genetic engineering possess defects in *hsd* methylation and problems are rarely encountered.

C *In vitro* methylation

The increasing availability of bacterial enzymes capable of methylating DNA at specific sequences has stimulated the use of these methylases as reagents for *in vitro* DNA modification. *In vitro* methylation of cloned DNA segments has been used to analyse the effect of methylation on gene expression *in vivo* (for instance Vardimon *et al.*, 1982; Waechter and Baserga, 1982). However, we will concern ourselves with DNA methylation as a tool for genetic engineering. As mentioned in section IV.A, *Eco*RI methylase has been used to protect internal sites in target DNA from scission at the last step of the linker addition procedure (Maniatis *et al.*, 1978; Kemp *et al.*, 1979). Such methylases also have the potential to be used to block specific sites while leaving others open to cleavage. For instance, the *Taq*I methylase could be used to protect specific *Acc*I or *Hin*dII sites from cleavage (McClelland, 1981b) and, as we have seen above, passage of DNA through *E. coli* can be used to specifically block a subset of *Cla*I sites. Methylation *in vitro* using the *dam* methylase (Geier and Modrich, 1979; Herman, 1981) could be a useful adjunct to avoid selective mismatch correction of a mutant strand during localised mutagenesis (see section VIII.D). A very useful tool for the molecular biologist would be a means to specifically remove methyl groups from DNA *in vitro* but, to our knowledge, no method exists for carrying this out.

VI Aberrant activities of restriction enzymes

A Relaxed specificity and site preference

It has been recognised for many years that restriction enzymes will relax their specificity under certain conditions. The best characterised case is the *Eco*RI* activity exhibited by *Eco*RI under conditions of low ionic strength (Polinsky *et al.*, 1975). The relaxation of site specificity is further enhanced by replacing Mg^{2+} in the restriction reaction by Mn^{2+} (Hsu and Berg, 1978). Although the canonical recognition sequence GAATTC is cleaved preferentially, sites differing in sequence at one or even two positions from the canonical sequence are also cleaved (Woodbury *et al.*, 1980; Gardner *et al.*, 1982) though not at the same rate. Kemp *et al.* (1979) successfully used the reduced specificity of *Eco*RI to construct a random segment bank of mouse DNA after methylating the DNA at canonical sites with the *Eco*RI methylase.

Major alterations in site specificity can also be provoked by the addition of organic solvents (Malyguine *et al.*, 1980) the best characterised example of which is the effect on cleavage by *Bam*HI (George *et al.*, 1980; George and Chirikjian, 1982). Again, the major effect is to reduce the selectivity of cleavage such that close relatives of the canonical sequence GGATCC are cleaved. A useful consequence of the different rates of cleavage of secondary sites is that alternative specific cuts may be made under suitable conditions. For instance, in the presence of 15% glycerol the major secondary cleavage site in plasmid pBR322 is about 1.3 kb from the *Bam*HI site, probably at the sequence GGATCT at nucleotide 1666 (Lathe, unpublished data) although this sequence occurs a number of times in the plasmid.

It should be noted that a differential rate of cleavage at secondary sites is not an isolated phenomenon, and even cleavage at the canonical sequence often varies considerably depending on the adjacent nucleotide sequences. For instance, *Pst*I exhibits a defined order of cleavage preference in either superhelical or linear DNA which is probably inversely related to the G/C content of the immediately flanking DNA sequences (Armstrong and Bauer, 1982). Site preference exhibited by most enzymes can be particularly inconvenient when specific partial digestions are to be performed. Selective inhibition of cleavage at particular sites can be further enhanced by the addition of intercalating drugs (for example, see Nosikov *et al.*, 1976).

B Nicking activities

In a number of cases, restriction enzymes will introduce site-specific nicks into DNA. In particular, a restriction enzyme may introduce a strand-specific nick in a sequence which differs in one position from the canonical sequence. Although *Eco*RI (recognition sequence GAATTC) has been reported to introduce a nick at the sequence

GAATTA (Bishop, 1979), little is known about the generality of this phenomenon. One consequence of this observation is that *Eco*RI should catalyze double-strand cleavage at positions where GAATTA is present in both strands, e.g. GAATTAATTC. This sequence is generated by self-joining a filled-in *Eco*RI terminus. In a second case, a site which is protected by methylation on only one strand (hemi-methylated) may nonetheless be nicked in the unprotected strand (for example *Sau*3A; Streeck, 1980). Note that mild restriction enzyme treatment of DNA in the presence of the intercalating agent ethidium bromide can be used to insert specific nicks at canonical sequences and this is of particular interest for localised mutagenesis (see later sections).

S1 nuclease (see section III.D) is capable of converting a single strand break into a double strand cleavage, and the combination of a nicking activity with this enzyme may enable specific ds cleavages to be performed upon stretches of DNA devoid of restriction sites. Further, nuclease S1 will itself introduce site-specific ds cleavage at certain inverted repeat sequences in superhelical DNA (Lilley, 1980, 1981).

VII Selection for the presence or absence of restriction sites

A Selection for absence

Murray and Murray (1974) plated λ phage upon a strain of *E. coli* producing restriction enzyme *Eco*RI in order to select for the loss of recognition sites. More simply, cutting with the restriction enzyme in question prior to transformation provides a sufficient selection for the absence of a site, although repeated cycles of growth and cleavage may be necessary (see Messing *et al.*, 1981). Also, recircularisation *in vivo* often occurs at a sufficient frequency to necessitate further manipulations; in particular dephosphorylation or S1 treatment after cleavage are useful adjuncts (Rosner *et al.*, 1978). Physical separation such as by CsCl-ethidium bromide centrifugation may also be employed to separate an uncleaved supercoil from a cleaved linear form.

B Selection for presence

To select for the presence of a restriction site in a population of molecules lacking that site is a difficult task. The easiest approach is, if possible, to identify a restriction site which is lost simultaneously during the alteration to gain a restriction site, and to select for the absence of the second site. For instance, the mutation in phage M13*mp*7 which removes the *Acc*I site simultaneously creates a unique site for *Bgl*II (Messing *et al.*, 1981; van Wezenbeek *et al.*, 1980).

```
   ******                  ******
TTTGTAGACCTCTCA  ----->  TTTGTAGATCTCTCA
     AccI site               BglII site
```

A different approach was used by Gronenborn and Messing (1978) to isolate a M13*mp*1 derivative containing an *Eco*RI site. They digested the population with *Eco*RI and enriched for linearised molecules by gel electrophoresis. After excision from the gel, recircularisation with DNA ligase and transformation, phages containing the *Eco*RI recognition sequence were isolated.

VIII Localised mutagenesis

The methods described in the preceding sections on restriction site manipulation can be considered to be straight-forward but limited approaches to directed mutagenesis. For example, the protruding 3′ extremities of the *Bgl*I site in SV40 were removed with S1 nuclease and ligated to show that this site is in a region essential for replication (Myers and Tijan, 1980). This type of experiment would often produce frameshift mutations within protein-coding regions and has also been used to analyse promoter function (for example, Grosveld *et al.*, 1982). However, the principles have been fully discussed above and since their utility for directed mutagenesis depends on an especially fortunate distribution of restriction sites, they will not be considered further.

In this section we will describe some of the most commonly used methods for producing deletion, insertion, transition and transversion mutants within a sequence. Often a first approach to the manipulation of a sequence of biological interest is the introduction of specific deletions; this aids the identification of segments for further study. Therefore, once this has been accomplished, the next step is often directed point mutagenesis to produce mutations within a given region or at a specific base. The choice of method depends to a certain extent on the sequence in question and the pattern of restriction sites within and outside that region. Otherwise the most successful methods are those which are adaptable, straightforward and give a low background of unmutagenised clones.

This topic has been recently reviewed in general by Shortle *et al.* (1981) while the methods of oligonucleotide mutagenesis were covered in depth by Smith and Gillam (1981). However, since these articles were published there has been an explosion both in new methods and in improvements to existing ones. We will attempt to compare the most recent methods with earlier approaches while stressing the basic requirements for the success of any given protocol.

A Deletion mutants

Deletion mutants may be produced by many different methods. These usually employ one or more restriction sites within a (generally circular) DNA molecule although some approaches do not require restriction sites. According to the protocol used the end-points of the deletions produced are either pre-defined or random. It is important to note that if deletion of a sequence causes a particular effect this may be due to altered spacing between sequences flanking the deletion rather than to loss of the nucleotide sequence itself.

1 Defined deletion mutants

The simplest case is where two homologous restriction sites exist within a circular molecule. Cleavage and re-ligation generates molecules with a defined deletion. Alternatively, site fusion of two different restriction termini (see section III) can be used. When more than two sites exist, ligation following a partial digest results in a family of deletions which have lost different defined segments of DNA. Such methods usually result in fairly extensive deletions and are rarely useful for localised mutagenesis. In contrast, an interesting method is that of Panayotatos and Truong (1981). They used a fragment of phage T7 containing the late promoter cloned in plasmid pBR322 between intact *Eco*RI and *Bam*HI sites. After cutting with *Bam*HI and limited digestion with *Bal*31 nuclease (see below), the molecules were cut with *Sal*I, the protruding ends filled-in and re-ligated. The resulting plasmids were screened for the presence of a *Sal*I site, arising from ligation of the filled-in *Sal*I terminus (5′ TCGAC—3′) to a 3′-terminal G on the *Bal*31-digested terminus. Deletion mutants were constructed that extended short distances to preselected bases, regenerating a useful restriction site at their end-points.

Panayotatos and Truong (1981) used an unique restriction site outside the range of the *Bal*31 for their experiments. However, it would be possible to produce very small defined deletions which otherwise retain all the sequences either side of the restriction site used by modifying their technique. For example, in the case described, the original plasmid could be cut with *Bam*HI, filled-in, cut with *Sal*I and the *Bam*HI-*Sal*I fragment purified. Ligation of this

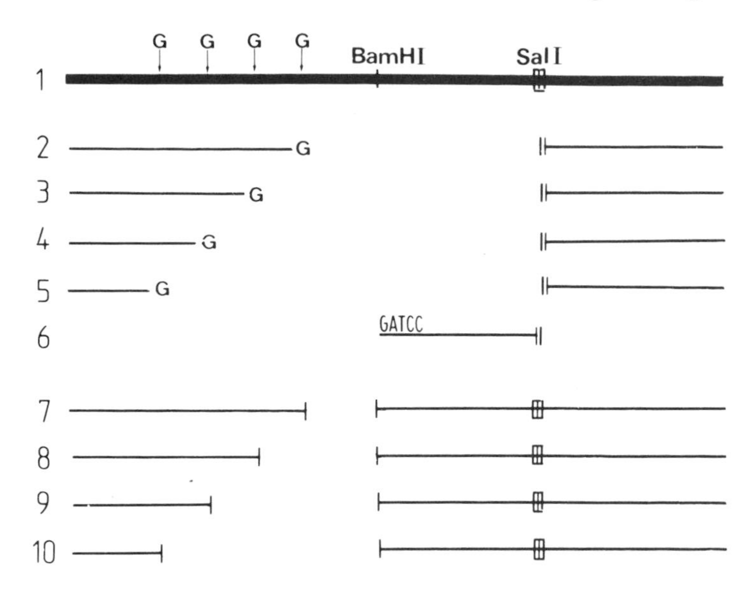

Figure 4 Deletion mutants extending from an unique point to pre-selected bases. 1, the original plasmid where the *Bam* HI and *Sal* I sites and the position of G residues to the left of the *Bam* HI site are marked; 2—5, linear molecules cut by *Bam* HI have been treated with *Bal* 31 nuclease, and then cut by *Sal* I. Some of the products will have the G residues shown as their end-point. 6, the filled-in *Bam* HI — cohesive *Sal* I fragment from 1; 7—10, fragment 6 ligated to molecules 2–5. The G residues at the end of each deletion combine with the GATCC sequence at the filled-in *Bam* HI site to produce a *Bam* HI site, indicated by the vertical lines. Assuming an equal distribution of bases, approximately 25% of the deletions produced will have *Bam* HI sites. These plasmids could be detected by screening either before or after selection for plasmids cut by *Bam* HI (see section VII).

fragment to the plasmid digested with *Bal* 31 at the *Bam* HI site, and subsequently cleaved with *Sal* I, regenerates *Bam* HI sites at each G-C as before. Such a protocol would have the advantage of leaving the original *Bam* HI-*Sal* I sequences intact (see Fig. 4).

These methods use unique restriction sites in circular DNA to produce defined deletions. However, it is possible to do analogous experiments with enzymes that cut the DNA at many positions provided that the enzyme does not cut ss DNA (see Wells *et al.*, 1981). Humayan and Chambers (1979) annealed a restriction fragment of ϕX174 ds-RF DNA to the ss phage DNA. The duplex portion of the heteroduplex contained 2 *Hinc*II sites; thus this segment could be excised, the molecule religated and mutant phages obtained after transfection. Potentially this method could be used to delete DNA between any two sites if the region to be studied can be cloned into a single-stranded DNA vector and suitable flanking restriction sites are available.

2 *Random deletion mutants*

We use the word random here to describe deletions whose end-points are not pre-defined by the protocol used. Nevertheless, the general region of the deletion is often selected because of a convenient, unique restriction site. The most common and effective method to produce random deletions has been to cut the DNA at an unique site and digest DNA from the ends using *Bal*31 nuclease. This enzyme acts both as an exonuclease and a single-strand specific endonuclease and therefore it will progressively digest DNA from exposed termini (Gray *et al.*, 1975). The use of *Bal*31 is more convenient than the use of ds-dependent exonucleases, such as exonuclease III ($3' \rightarrow 5'$) or λ-exonuclease ($5' \rightarrow 3'$) since these enzymes leave exposed ss DNA at the termini of the DNA which must be removed by the additional step of S1 nuclease treatment before re-ligation. Unfortunately all of these enzymes may pause at certain DNA sequences giving rise to preferred deletion end-points. These end-points differ according to the enzyme used. In practice it is often advantageous to treat the exonuclease digested DNA with polymerase I in the presence of the four dNTPs to make precisely blunt all those termini left ragged by this treatment (see section III).

The use of linkers at the ligation step to generate an unique site in the deleted molecule is especially recommended, since this facilitates mapping and subsequent manipulations. For example, Fig. 5 shows how it is possible to create a complete set of overlapping deletions by making two sets of deletions progressing towards each other from restriction sites bounding the relevant region, an approach which has been used in several recent studies (for example, Tyndall *et al.*, 1981). This bi-directional protocol can be used to produce deletions within an area where no convenient sites exist (see Fig. 5). Deletions have been made in the 21 bp repeats of the SV40 early promoter region by using a deletion originating from a site 114 bp downstream and others from a site 187 bp upstream of the region of interest (Everett, R.D. and Chambon, P., in preparation). McKnight and Kingsbury (1982) have also used this approach.

Linkers also allow so-called "clean deletions" to be produced, where one end-point is constant while the other extends progressively into the desired sequence. These are obtained by using a linker oligonucleotide with the recognition sequence for the same enzyme which cuts the unique site starting point, or by suitably converting this site with a linker beforehand. Combining the deletions with the

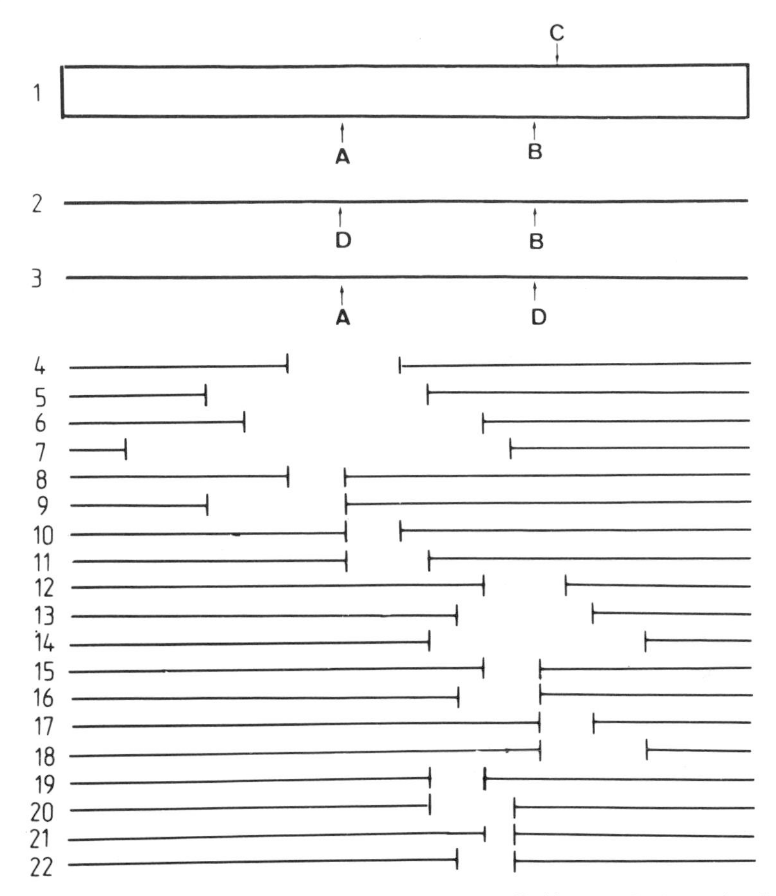

Figure 5 Deletion protocols using unique sites, *Bal*31 and linkers. 1, plasmid with unique sites A, B and C. The targets for deletion are at site A, at site B or between these sites. On subsequent lines, only that part of the plasmid which contains sites A and B is shown; 2, related plasmid where site A has been modified to unique site D with a linker; 3, site B has been modified to site D. Plasmids 4—7, deletions from site A. The extent of the deletion is shown by the blank area bounded by vertical lines which represent the insertion of a linker of site D specificity. Plasmids 8, 9, "clean deletions" extending leftwards from site A created by ligation of the left-hand-sides (lhs) of deletions 4 and 5 with the right-hand-side (rhs) of plasmid 2 using enzymes C and D. Plasmids 10, 11, "clean" deletions extending rightwards from site A using lhs 2 and rhs 4 and 5; plasmids 12—14, deletions from site B. Plasmids 15, 16, "clean deletions" from site B using rhs 3 and lhs 12, 13 and enzymes C and D. Plasmids 17, 18, deletions using lhs 3 and rhs 12, 13. Plasmids 19—23, deletions between sites A and B. Plasmids 19, 20 contain lhs 14 plus rhs 6, 7 respectively. Plasmids 21, 22 contain rhs 7 plus lhs 15, 16. These last four examples show how it is possible to create families of "clean" deletions starting from the end-point of any particular deletion mutant once the two sets of mutants have been obtained. Note also that combinations of, for example lhs 12 and rhs 10 give tandem duplications of the sequences between the two deletion end points.

undeleted plasmid produces variants which have a fixed deletion end-point (Fig. 5). It is often easier to interpret biological results obtained with this type of construction since only one deletion end-point varies.

This strategy can be extended to non-unique sites using the single-strand hybridisation method discussed earlier for deletion between non-unique restriction sites (Lambert, P. F. and Wells, R. D., personal communication).

Methods have been described where deletions are introduced into plasmids in regions where no convenient sites exist. The first uses partial DNase I digestion in the presence of Mn^{2+} to produce random double-stranded breaks in the DNA (Shenk *et al.*, 1976). The DNA termini were made blunt with DNA polymerase I, and then ligated to incorporate *Eco*RI linkers. This generated plasmids with single randomly distributed linker insertion mutations (Heffron *et al.*, 1979). Combinations of such plasmids in a similar fashion to that shown in Fig. 5 could then be used to generate deletion mutants regardless of the existence of natural restriction sites. The disadvantage of this method is that the whole plasmid is attacked in the first step; if the region of interest is comparatively small, very many mutants would have to be analysed to find those required.

The approach of Green and Tibbetts (1980), although more prone to experimental difficulties, does not suffer from this problem. They used an isolated single-stranded fragment of DNA to displace one strand of a closed-circular plasmid molecule (D-loop formation). The efficiency of this step can be increased using *E. coli rec*A protein (Shibata *et al.*, 1979; McEntee *et al.*, 1979). The exposed single-stranded region could be nicked by S1 nuclease, leading to the collapse of the D-loop (Fig. 6), and under their conditions, a double-stranded cut with loss of a small number of nucleotides. The purified linear molecules were ligated and mutants with small deletions in the target area obtained. Such mutants are more useful for many experiments if linkers are used during ligation since the linker insertion mutations can be used to produce a series of second generation deletion mutants.

B Region-directed point mutagenesis

The introduction of point mutations into a target region, but at random within that region, is a very powerful approach. Many methods are available, some using chemical mutagenesis *in vitro*, others tools such as error-prone polymerases or nucleotide analogues. The mutations produced are either transitions (which interchange pyrimidine bases or purine bases, e.g., C → T or G → A) or

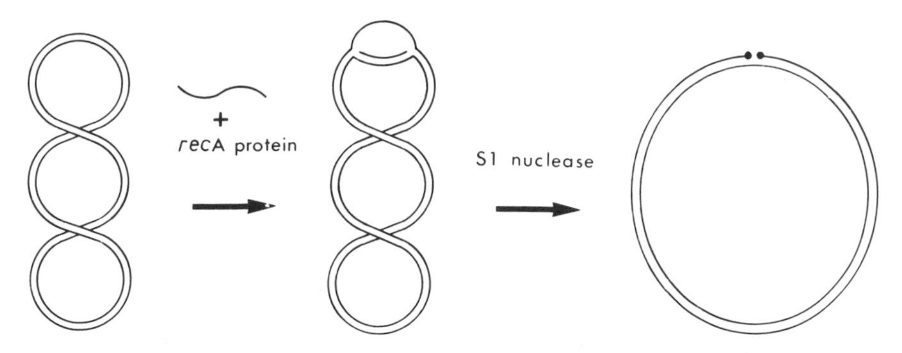

Figure 6 Creation and destruction of D-loop molecules. Supercoiled circular DNA is incubated with an isolated single-stranded fragment complementary to a sequence in the plasmid. The *E. coli recA* protein catalyses a reaction where the single-strand fragment is paired with its complementary sequence in the circular molecule, leading to displacement of the other strand in a small single-stranded loop (D-loop). Due to the torsional stress of this operation, it can only occur with covalently closed circular molecules. Therefore, on incubation with S1 nuclease, the D-loop collapses immediately once the first exposed single-strand phosphodiester bond is broken. Under certain conditions of S1 nuclease treatment, the enzyme may go on to produce a double-strand break. These nicks or breaks will only occur in the exposed D-loop region, so that the target area can be varied according to the single-stranded fragment used.

transversions which exchange pyrimidines with purines (e.g., G → T). Unlike deletion mutants, point mutants are rarely open to analysis using restriction enzymes. Methods using single-stranded DNA bacteriophage vectors such as M13 and fd have a particular advantage since the sequence of the mutant region can often be rapidly established using dideoxy chain termination sequencing (Sanger *et al.*, 1980). Thus large numbers of mutants can be screened before choosing those most suitable for further study.

1 Isolated fragment mutagenesis

The basic protocol here is to isolate the desired region as a restriction fragment and mutagenise *in vitro* with hydroxylamine (which deaminates cytosine bases to hydroxyaminocytosine, a thymine analogue) or nitrous acid (which deaminates adenine to hypoxanthine (a guanine analogue) and cytosine to uracil; Zimmerman, 1977). The mutant fragments can then be re-cloned into a suitable vector. This method has proved powerful in instances where the mutant fragment can be reintroduced into its natural location either *in vitro* or by recombination *in vivo* and where mutant phenotypes are easily recognised (Volker and Showe, 1980; Solnick, 1981; Busby *et al.*, 1982; Conley *et al.*, 1981; Sandri-Goldin *et al.*, 1981). However, the yield of mutants may be poor, they are not amenable to rapid sequencing and only those liable to give a recognisable change in phenotype will be detected. It is important to realise that mutations which show no altered phenotype may provide interesting biological information.

2 *Mutagenesis with sodium bisulphite*

Sodium bisulphite deaminates cytosine bases to give uracil, but only in single-stranded DNA (Kai *et al.*, 1974). Therefore all the protocols using bisulphite involve the production of a defined ss region *in vitro*. The simplest case is at the protruding ends of a restriction enzyme site. It is often straightforward to destroy restriction sites within the drug resistance genes of plasmid vectors and produce variants that still express resistance. For example, after bisulphite mutagenesis and religation of *Bam* HI-cut pBR322, a tetracycline and *Bam* HI resistant clone was obtained (Everett, R. D., unpublished). The degree of mutation within any region exposed to bisulphite can be varied by choice of the concentration of the mutagen and the incubation time used.

(*a*) *The use of double-stranded vectors.* The first published mutagenesis experiments using sodium bisulphite employed double-stranded vectors. For example, Shortle and Nathans (1978, 1979) nicked SV40 DNA with *Bgl*I in the presence of ethidium bromide, produced a small gap with *Micrococcus luteus* DNA polymerase, and mutated the exposed cytosines with sodium bisulphite. The method was extended by DiMaio and Nathans (1980), who first "translated" a nick at the same site in SV40 with *E. coli* DNA polymerase I before gap generation and mutagenesis. Mutants were selected by altered plaque phenotype, cold sensitivity or restriction site alterations.

These experiments limit their target region to an area close to a suitable restriction site. However, Shortle *et al.* (1980) used the approach of Green and Tibbetts (1980) described earlier to produce a D-Loop within the β-lactamase gene of pBR322 (see Fig. 6). Supercoiled molecules with D-loops were incubated with S1 nuclease under conditions where the exposed single-stranded DNA in the D-loop was nicked, leading to collapse of the D-loop and production of a nicked open-circular DNA molecule. On treatment with *Micrococcus luteus* DNA polymerase I in the absence of dNTPs, the nicks were converted to small gaps which were then treated with bisulphite. Finally the mutagenised gaps were repaired with DNA polymerase I and any supercoiled DNA remaining removed (by acridine yellow chromatography) before transformation of *E. coli.*

Mutant clones, sensitive to ampicillin, could be easily identified. This method has the advantage that the target area can be defined by any two restriction sites. However, it is technically complex and likely to generate a high background of un-mutagenised clones. Therefore it is probably only applicable in cases where mutants can be easily identified.

A further method for bisulphite mutagenesis using double-stranded vectors has been described by Giza *et al.* (1981). They produced mutations in the tetracycline resistance gene of pBR322 between the *Bam* HI and *Sal* I sites. The desired single-stranded gap was prepared by annealing an isolated single strand from *Hpa* I-cut pBR322 (*Hpa*I cuts the plasmid only once) with the complementary single strand from the plasmid cut by both *Bam* HI and *Sal* I. Heteroduplexes contained an exposed ss region between the *Bam* HI and *Sal* I sites which was treated with bisulphite, and filled-in before transformation. Although technically simpler than the method of Shortle *et al.* (1980) this method has the disadvantage that prior preparation and identification of isolated single-strands from the whole plasmid is required and is also limited by the requirement for unique sites bordering the region of mutagenesis.

The simplest bisulphite mutagenesis protocol for double-stranded vectors described so far is that of Kalderon *et al.* (1982). They used a two-step procedure that uses previously isolated deletion mutants in the cloned SV40 T-antigen gene. They cut these molecules at a unique restriction site and made heteroduplexes with the equivalent "wild-type" clone (without the deletion) cleaved at a different unique site. Circular heteroduplexes contained a loop-out of single stranded DNA in the "wild-type" strand at the position of the deletion in the mutant strand. Homoduplexes (also formed during the annealing reaction) were linear. The mixture was treated with bisulphite and transformed directly into an *E. coli* strain lacking the enzyme uracil-N-glycosylase (*ung*$^-$) (see below). Since circular heteroduplexes transform the recipient at a much higher frequency than linear homoduplexes, one of the principal sources of "wild-type" background is eliminated. One might expect that the clones would contain a mixed population of the multicopy plasmid, including molecules both with and without the original deletion, the two types being derived from the strands of the parent heteroduplex. In practice, however, most of the clones contained only one of the 2 types perhaps because of mis-match repair before replication. Those plasmids of wild-type size were screened by DNA sequence analysis, and several different mutants identified. This method fulfils many criteria for a convenient directed mutagenesis protocol. It is simple, rapid, requires no special materials or techniques and results in a high yield of mutants. The use of previously isolated deletion mutants eliminates the need for convenient restriction sites. However, in common with other methods using ds vectors, the most convenient methods of rapid DNA sequence analysis are not applicable.

The "deletion-loop" mutagenesis method of Kalderon *et al.* (1982) uses an *E. coli* (*ung*$^-$) recipient strain because otherwise the enzyme uracil-N-glycosylase excises the uracil residues so eliminating the mutants (Tye *et al.*, 1978). Unfortunately, this strain takes up exogenous DNA at a frequency 10-fold lower than the more commonly used recipients. The use of this strain is unnecessary in procedures that include a filling-in step, as adenosine nucleotides are incorporated opposite the uracils *in vitro*, ensuring the inheritance of the mutation.

(*b*) *The use of single-stranded vectors.* The use of ss DNA bacteriophage vectors, such as M13 and fd, during bisulphite mutagenesis procedures has two immediate advantages. First the construction of a defined ss region is often technically simple and secondly the resulting clones are easily screened by rapid DNA sequence analysis. M13 vectors are particularly recommended because of the large number of available variants with different restriction sites for introducing the fragment to be mutated (Messing, 1982), and the readily available primers for DNA sequencing (Sanger *et al.*, 1980).

Published procedures using ss vectors often use similar strategies to those employing ds vectors, but make use of one or more of the particular advantages noted above. Weiher and Schaller (1982) cloned a fragment of the *lacUV5* promoter into phage fd, mutagenised the whole viral single strand with bisulphite and then used a suitable primer to produce ds DNA in the region of the cloned fragment. The mixture was then digested with suitable enzymes to allow re-cloning of the mutagenised fragments back into a plasmid vector. This simple method ignores the advantages of rapid sequencing in ss phage since re-cloning into a plasmid vector was performed before the mutant phage could be isolated and characterised. Ciampi *et al.* (1982) used rapid sequencing in M13 to analyse mutations within the tRNA^{Pro} gene of *Caenorhabditis elegans.* However, they produced the exposed single-stranded region necessary for bisulphite treatment by DNA polymerase I 3′—5′ exonuclease action on the *Sma*I cut replicative (ds) form of the phage.

Everett and Chambon (1982) made use of the advantages of ss vectors in a two-step procedure to isolate mutations within the 21 bp repeat region of the SV40 early promoter. Deletion mutants had been produced within this region using *Bal*31 and *Bam*HI linkers as described earlier. The "wild type" and deletion mutant early promoter fragments were then cloned into M13*tg*103 and heteroduplexes formed between the ss viral DNA of the wild-type clone and the RF of a deletion mutant cleaved by the single-cutting *Bam*HI. Thus the heteroduplexes contained a single-stranded region corresponding to the extent of the original deletion. After bisulphite mutagenesis, filling-in and transfection the resulting phages were screened directly by DNA sequencing. Except when the mutagenesis conditions used were exceptionally mild, all phages contained mutations within the desired region. This high efficiency arises since the whole genome of unhybridised ss circular vector DNA is exposed to bisulphite and therefore is mutated, in essential genes, and also any linear homoduplexes of the *Bam*HI-cut deletion mutant give a low frequency of transfection. Therefore it is only the mutated heteroduplexes which give rise to plaques at high frequency. This method can be used when deletion mutations are not available by cutting the RF DNA to excise a small fragment before heteroduplex formation, or cloning the target sequence by itself and using the cleaved vector RF to protect the phage sequences from mutation.

(*c*) *Directed mutagenesis in vivo.* An interesting adaptation of techniques using ss phage was described by Traboni *et al.* (1982). Rather than using bisulphite as a mutagen, these workers directed mutations into the cloned $tRNA^{Pro}$ gene of *Caenorhabditis elegans* during phage growth using traditional mutagens for *in vivo* mutagenesis such as ICR191, aminopurine or nitrosoguanidine (Miller, 1972). This was possible because the $tRNA^{Pro}$ gene was inserted into the N-terminal segment of the β-galactosidase gene in M13*mp*701 (in which the cloning sites lie) such as to leave the reading frame intact. The inserted fragment contained no translational stop codons. Thus the recombinant phage gave blue (*lac*$^+$) plaques on suitable indicator plates. (Traboni *et al.*, 1982, give a description of M13 virus/host interaction which allows the production of β-galactosidase in cells infected with vector virus, but not normally with those containing an inserted DNA fragment.) Following mutagenesis during growth of the phage *in vivo*, Traboni *et al.* (1982) obtained a number of white (*lac*$^-$) plaques, some of which contained single point mutations in the $tRNA^{Pro}$ gene. These were either frameshifts (induced by ICR191) or translational stop signals (induced by nitrosoguanidine or aminopurine). This method is very simple and should be applicable to many protein-coding DNA fragments which can be cloned in phase.

C Mutagenesis using nucleotide analogues, mis-incorporation, and error-prone polymerases

1 Nucleotide analogues

Most of the methods described above generally give multiple or random mutations within a target region, whereas nucleotide analogues sometimes allow a particular base to be altered. The analogues used for mutagenesis are nucleoside triphosphates which can be incorporated during synthesis by DNA or RNA polymerases but, due to alternative possible structures for hydrogen bonding, may be paired with either of two other bases. For example, N^4-hydroxy dCTP has two tautomeric forms; the amino form pairs with guanine whereas the imino form pairs with adenine. Similarly 2-aminopurine deoxyriboside triphosphate may pair with either pyrimidine base according to its tautomeric form (Watanabe and Goodman, 1981). The basic requirement for an experiment using nucleotide analogues is a primer for *in vitro* DNA synthesis positioned appropriately on a template strand close to the site to be mutagenised. It is interesting that the incorporation of a nucleotide analogue during *in vitro* synthesis is not sufficient to give rise to a mutation. Rather, the mutagenic event occurs during a later replication step (often *in vivo*) while the analogue is in its alternative tautomeric form. Site-directed mutations have been produced in the genome of the single-stranded bacteriophage Qβ using N^4-hydroxy dCTP, the Qβ replicase and the viral RNA which self-primes its own replication (Flavell *et al.*, 1974; Weissmann *et al.*, 1979). Muller *et al.* (1978) used an unique nick introduced at a restriction site as a primer for DNA synthesis in the presence of N^4-hydroxy dCTP, and Weber *et al.* (1981) used isolated primer fragments hybridised to ss phage vectors.

The exact protocol depends on the base to be mutagenised and the surrounding sequence. Figure 7 illustrates the basic principles of an experiment using N^4-hydroxy dCTP. The adenine at position 22 is the target for mutation. The initial step is to elongate the primer to within reach of the target. This can be done in two stages, first in a reaction which lacks dGTP (so that synthesis stops at base 18) and secondly, after ethanol precipitation and extensive washing of the DNA, in a reaction lacking dTTP. This stops the polymerase at base 21. The next step is to wash the DNA again and incubate with polymerase, N^4-hydroxy dCTP and dCTP. The analogue will be incorporated at base 22 in its imino form, while the presence of dCTP competitively inhibits its incorporation in its amino form at position 23. Thus the mutants will be A $\rightarrow$ G transitions at position 22. Alternatively, if dTTP were added instead of dCTP, the analogue would not be incorporated at position 22 because of competition

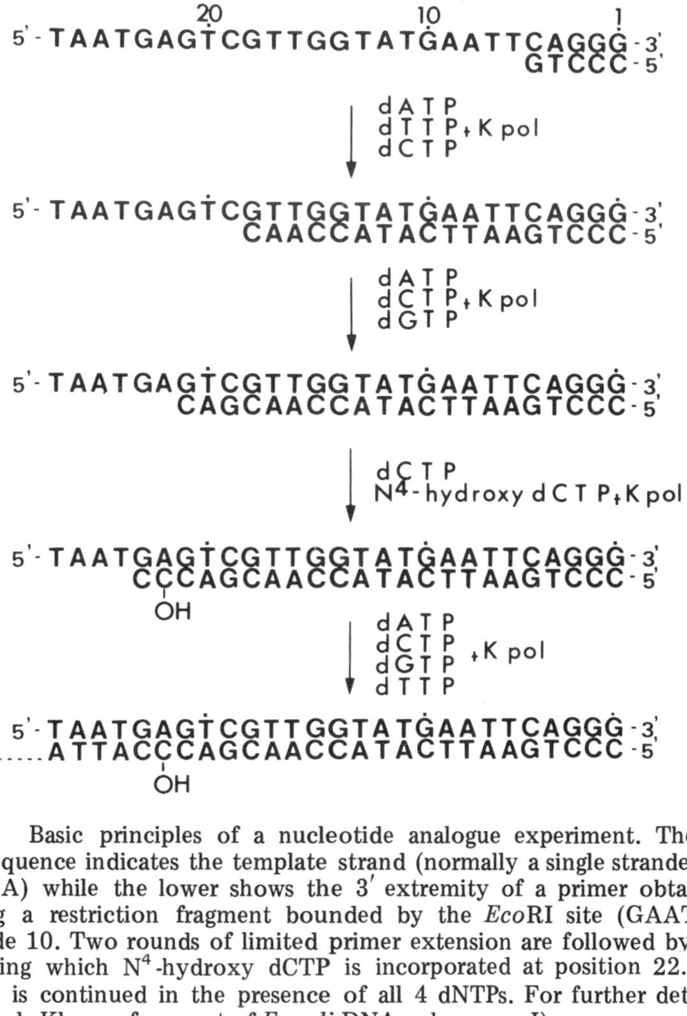

20 10 1
5'-TAATGAGTCGTTGGTATGAATTCAGGG-3'
GTCCC-5'

dATP
dTTP + K pol
dCTP

5'-TAATGAGTCGTTGGTATGAATTCAGGG-3'
CAACCATACTTAAGTCCC-5'

dATP
dCTP + K pol
dGTP

5'-TAATGAGTCGTTGGTATGAATTCAGGG-3'
CAGCAACCATACTTAAGTCCC-5'

dCTP
N4-hydroxy dCTP + K pol

5'-TAATGAGTCGTTGGTATGAATTCAGGG-3'
CCCAGCAACCATACTTAAGTCCC-5'
OH

dATP
dCTP
dGTP + K pol
dTTP

5'-TAATGAGTCGTTGGTATGAATTCAGGG-3'
.....ATTACCCAGCAACCATACTTAAGTCCC-5'
OH

Figure 7 Basic principles of a nucleotide analogue experiment. The upper strand sequence indicates the template strand (normally a single stranded phage viral DNA) while the lower shows the 3′ extremity of a primer obtained by preparing a restriction fragment bounded by the *Eco*RI site (GAATTC) at nucleotide 10. Two rounds of limited primer extension are followed by a third step during which N^4-hydroxy dCTP is incorporated at position 22. Finally synthesis is continued in the presence of all 4 dNTPs. For further details, see text. (Kpol: Klenow fragment of *E. coli* DNA polymerase I).

from the dTTP, but at position 23 to allow the formation of G → A transitions at that base instead. Finally chain elongation is completed in the presence of all four dNTPs. The hydroxycytosine base at position 22 will have either the adenine or guanine paired with it during later replication, depending on its tautomeric form at the time. Therefore some of the progeny molecules will be mutant specifically at that position. In principle almost any base can be mutated in this fashion, but in practice only those within a few nucleotides of the original primer are easily accessible since repeated prior elongation steps lead to experimental problems.

In the work of Weber *et al.* (1981) the elongated, mutated primer strand was purified from the template using alkaline sucrose gradient centrifugation, made double-stranded with DNA polymerase (either by self-priming or by adding a complementary strand primer), then cut with appropriate restriction enzymes and re-inserted into a plasmid vector. This complicated procedure could probably be simplified by allowing the second strand synthesis to proceed to complete a ds replicative form circle for transfection; mutants could then be identified by rapid dideoxy sequencing.

This approach could be used for region-directed point mutagenesis in conjunction with any of the protocols described above for bisulphite mutagenesis which require a filling-in step. By using N^4-hydroxy dCTP in the place of dCTP or dTTP during the filling-in one could generate a number of A → G or G → A transitions (in the exposed single-strand) instead of the C → T transitions produced by bisulphite. The degree of mutagenesis in experiments where all 4 normal dNTPs as well as N^4-hydroxy dCTP were present can be controlled by suitable adjustment of the concentrations of dCTP, dTTP and N^4-hydroxy dCTP.

2 *Mis-incorporation and error-prone polymerases*

Under certain conditions of complementary strand synthesis by DNA polymerases, the enzyme can incorporate an incorrect nucleotide, thus producing a point mutation in the newly synthesized strand (see Kornberg, 1980; Kunkel and Loeb, 1979). In most cases this nucleotide is removed by the 3′—5′ "proof-reading" exonuclease of bacterial DNA polymerases. However, 3 methods have been described to avoid this problem. The first 2 use gapped circular DNA as a template for mutagenesis. These gaps were produced by *Micrococcus luteus* DNA polymerase I activity on templates nicked by the action of a restriction enzyme in the presence of ethidium bromide (Shortle *et al.*, 1982). In the first method the gap was filled-in in the presence of only 3 of the 4 dNTPs plus an excess of DNA ligase. More than 90% of the gapped templates were converted to covalently-closed circles during extended incubation and therefore one of the three nucleotides present must have been incorporated in place of the missing nucleotide at all relevant positions. Between 6 and 33% of the transformants analysed contained either transition or transversion mutations (Shortle *et al.*, 1982). In a second method, α-thiophosphate nucleoside triphosphates were used. Once incorporated, these nucleotides are resistant to excision by the 3′—5′ exonuclease activity of bacterial DNA polymerases, allowing mis-incorporation events to be stably inherited. Here the gapped templates were incubated first with DNA polymerase Klenow fragment and the α-thiophosphate (either α-SdTTP or α-SdATP in the absence of other dNTPs) to produce mis-incorporation at the 3′-OH primer terminus. After filling-in completely with all four dNTPs, 13–42% of plasmids isolated after transformation had transition or transversion mutations at the primer terminus (Shortle *et al.*, 1982). Both these approaches give rise to a fairly high proportion of deletion mutants which may result from the ability of T4 DNA ligase to seal small single-stranded gaps (Nilsson and Magnusson, 1982). The protocols limit the target area to within a few nucleotides of a unique restriction site, but in principle the method could be extended to produce mutations in the region of any restriction site using single-stranded phage and primer systems.

A third approach to mis-incorporation is to use an error-prone polymerase. DNA polymerase ("reverse transcriptase") from avian myeloblastosis virus (AMV) lacks the 3′–5′ "proof-reading" exonuclease (Battula and Loeb, 1976). Therefore mistakes made during polymerisation are not corrected. Zakour and Loeb (1982) annealed a primer fragment to φX174 viral single-stranded DNA and used this material for limited (faithful) synthesis with *E. coli* DNA polymerase I to extend the primer to the desired position (as shown in Fig. 7). The elongated primer-template was then incubated with an incomplete set of dNTPs and AMV DNA polymerase; again, incorrect nucleotides were incorporated where the correct one was not available. Finally, complementary strand synthesis was completed in the presence of all 4 dNTPs. Using this method up to 33% reversion of an amber φX174 mutant was obtained by site-directed transition or transversion mutagenesis.

All three methods can be used to mutate pre-selected bases, but only when using α-thiophosphate nucleotides can the replacement for the target base be specified.

D Oligonucleotide-directed mutagenesis

This method has recently been reviewed in depth by Smith and Gillam (1981). Therefore the principles and technical details will be considered rather briefly here. Despite difficulties in obtaining oligonucleotides of defined sequence (although some are now commercially available) it should be realised that in many instances this approach provides the only realistic method for obtaining a desired mutation.

The basic procedure is that a 5′-phosphorylated oligonucleotide containing the required mutation is hybridised to a circular ss template and used as a primer for elongation by *E. coli* DNA polymerase I Klenow fragment. The presence of T4 DNA ligase and ATP allows covalently closed circles to be formed on completion of the complementary strand and therefore prevents loss of the mutation by strand displacement of the primer region during a second round of synthesis. Molecules which have not been converted to full ds circles, either because of unsuccessful priming or incomplete elongation, are removed by limited S1 digestion before transfection. Plaques are then screened by phenotype, creation of a restriction site (often the easiest method), direct DNA sequencing, or differential hybridisation.

The basic requirements of the method, are first, that the fragment containing the site to be mutagenised be cloned into a single-stranded DNA vector such as ϕX174, fd or M13, and secondly that the oligonucleotide should be complementary to the wild-type sequence except for the mis-match which produces the desired mutation. In some instances two-base mismatches have been successfully used (Wasylyk, B., personal communication) and deletions of one base (Gillam *et al.*, 1980), three bases (Miyada *et al.*, 1982) and even 14 bases (Wallace *et al.*, 1980) have been obtained.

The important experimental criteria are the length and sequence of the oligonucleotide primer and its degree of mis-match with the template. The primer must hybridise to the required site and no other to avoid unwanted priming events; therefore, preferably, the complete sequence of the vector and insert should be known so that an oligonucleotide which will hybridise to the desired region only can be made. This means in most cases that the primer should be at least 7 nucleotides long but in practice oligomers of 10 to 12 bases have usually been used to produce single point mutations. Where double mutations or deletions are required even longer primers are needed; Wallace *et al.* (1980) used a 21 base primer to make a 14 bp deletion of the intron in a yeast tRNA gene. The position of the mis-match should be 3 or more bases from the 3′ end of the primer to avoid proof-reading excision by the Klenow polymerase (Smith and Gillam, 1981). Both hybridisation of the primer to the template and complementary strand synthesis were found to be more efficient if the incubation was performed at 0°C. Also, the decreased activity of the 3′—5′ "proof-reading" exonuclease activity of the polymerase at this temperature resulted in a higher recovery of mutants.

Isolation of the mutant progeny from plaques obtained after transfection is often not easy. Problems during the *in vitro* steps, such as excision repair of the mis-match by the Klenow polymerase, and mis-match repair *in vivo*, often result in a low yield of mutants, sometimes less than 1%. Two screening methods have been described, which rely on the fact that the oligonucleotide used for the original priming event will hybridise more strongly to the mutant viral strand than to the wild type. The first is simply plaque hybridisation; under suitable conditions mutant plaques hybridise more efficiently to the labelled oligonucleotide used for mutagenesis than will the wild type (Wallace *et al.*, 1980). The second method uses sequential rounds of priming with the oligonucleotide. After the initial *in vitro* reactions and transfection, progeny phage DNA is isolated and a further *in vitro* reaction performed with the same primer. Since the primer hybridises preferentially to the mutant phage DNA, the progeny phage from this second priming are enriched for the mutant. After three such cycles of enrichment Gillam *et al.* (1980) obtained mutant plaques at a frequency of 100%.

Recently, Kramer *et al.* (1982) have described some modifications to the basic protocol which lead to a greatly improved frequency of mutant isolation. They hybridised the M13 vector phage RF-DNA which had been cut with a restriction enzyme at the cloning site(s) to the viral ss of the M13 phage with a cloned insert to form a gapped duplex molecule. Subsequently an oligonucleotide primer was hybridised within the gap. This approach minimises non-specific priming because only the ss area is available for hybridisation. The region to be filled-in is also greatly reduced, leading to a much higher frequency of duplex circle formation. Furthermore, they were able to greatly reduce the level of mis-match repair *in vivo* by ensuring that the strand containing the mutation was methylated at the N^6 adenine position in GATC sequences (see section V.B). The basis of this approach is that heteroduplexes between genetically distinguishable λ phages, one methylated at GATC sites while the other is not, yield progeny containing genetic markers predominantly from the methylated strand: mismatch repair *in vivo* seems to show a preference for editing the unmethylated strand (Radman *et al.*, 1980). Thus in classical oligonucleotide mutagenesis experiments using filamentous phage vectors, it is the wild-type strand that is methylated while the required mutant strand made *in vitro* is not. This leads to preferential loss of the mutation *in vivo*. Kramer *et al.* (1982) showed that the frequency of marker yield using filamentous phage vectors could be dramatically increased by making heteroduplexes between methylated linear RF-vector DNA and unmethylated recombinant phage ss DNA prior to addition of the mutant oligonucleotide (see above). In this case the mutation lies in the methylated strand of a hemi-methylated duplex and so it is the mutant strand which is preferentially conserved. The unmethylated single-strand DNA was obtained by growth of the phage in an adenine methylase deficient (*dam*$^-$) *E. coli*, while the methylated replicative form DNA was isolated from a *dam*$^+$ host. In fact it is fortunate that the original oligonucleotide mutagenesis experiments used ϕX174, since this phage lacks GATC sequences and the problem of *in vivo* repair was not pronounced.

Recently Wallace *et al.* (1980, 1981) used oligonucleotides to introduce mutations into plasmid vectors. Their approach was to nick the plasmid DNA (at a restriction site using the enzyme in the presence of ethidium bromide), and to remove the nicked strand by extensive digestion with exonuclease III. This leaves a circular ss molecule which can be used for hybridisation with the oligonucleotide as before. However, it is liable to be more difficult to obtain clean template DNA by this method, and although it would sometimes avoid cloning steps, the use of single-strand vectors is in general to be preferred.

E Allele replacement

The protocols described above for the creation of site or region specific point mutations and deletions have, for the most part, used systems where small fragments of DNA are cloned into plasmid or single-stranded phage vectors. In some cases, for example, if *in vitro* transcription is being analysed (see Traboni *et al.*, 1982; Wasylyk *et al.*, 1980) the characterized mutants obtained can be studied immediately since the required experiments can be conducted on isolated restriction fragments. In many cases, however, it may be desirable to return the mutation to its normal genomic environment. For example, if a specific mutation had been produced in part of a protein coding sequence it would be necessary to exchange the mutant with the corresponding wild-type sequence in a construction containing the whole gene before the effect of the mutation on protein structure or function could be studied. This should be borne in mind before the experiment as a whole is designed. In the simplest cases, where bacterial plasmid or small circular DNA virus genes or sequences are being studied, re-cloning of the mutant sequence into its normal background would probably be quite straightforward using *in vitro* recombination methods perhaps coupled with the "sabotage" strategies discussed above (Section III).

As the complexity of the genome increases so the problem becomes more acute, and recovery of the mutant sequence becomes less efficient. However, Kapoor and Chinnadurai (1981) have described a technique that should be applicable to many linear DNA viruses of intermediate size. They co-transfected a small restriction fragment from the middle of the genome of the eukaryotic virus Adenovirus-2 (which contained a mutation constructed *in vitro*) with viral segments from the left and right ends of the genome which both overlapped the central fragment but not each other. Recombination *in vivo* in the overlapping regions gave rise to viable virus genomes. In still more complex systems this approach would be difficult because multiple restriction sites would make preparation of the required fragments difficult. However, in some cases, for example, the Herpes viruses, exceptionally efficient recombination *in vivo* gives an acceptable recovery of mutant viruses if a mutated restriction fragment is co-transfected with wild-type viral DNA (Conley *et al.*, 1981; Sandri-Goldin *et al.*, 1981).

The reintroduction of mutations made *in vitro* into their correct locations in the chromosomes of prokaryotic and eukaryotic cells relies on *in vivo* homologous recombination. Extensive knowledge of the genetics of bacteria and their plasmids and phages often allows the use of genetic tricks for the reintroduction of a mutation back into its proper chromosomal location (for example, Ruvkun and Ausubel, 1981). Similarly, Scherer and Davis (1979) have described a method for the exchange of the wild-type *his*-3 gene with a mutant allele in the simple eukaryote *Saccharomyces cerevisiae*. However, the reintroduction of mutations made *in vitro* into their proper chromosomal locations in higher eukaryotic cells has not yet been demonstrated.

Exogenous DNA can be transferred into many different eukaryotic cell lines using transfection with polycations such as DEAE-Dextran (Pagano, 1970) or co-precipitation with calcium phosphate (Wigler *et al.*, 1977) or physical techniques such as micro-injection (Graessmann and Graessmann, 1976) or protoplast fusion (Schaffner, 1980). Retrovirus vectors for the biological transfer of DNA into cells are now available (Doehmer *et al.*, 1982; Taboni *et al.*, 1982). However, in all these cases, when the exogenous DNA is stably inherited into the host chromosomes, it is probably at random locations and often in multiple or rearranged copies (see Anderson *et al.*, 1982; Scangos *et al.*, 1981; Scangos and Ruddle, 1981 and references therein).

Thus a complete gene with its own promoter and other important regulatory sequences may be transferred into higher eukaryotic chromosomes but, in contrast to the case in yeast, is rarely integrated by homologous recombination. This can present two problems. First, the study of the mutant gene may be complicated by the presence of the resident wild-type genes, and secondly the behaviour of a gene may be affected by its chromosomal location. At present there would appear to be little one can do to improve this situation (although for some experiments these considerations may not be important). However it may be worthwhile to consider the use of cosmid vectors (Collins and Hohn, 1979). The mutation could be transferred *in vitro* from its initial vector to a cosmid vector, containing as large a region as possible of its normal genomic flanking sequences, and then transfected into cells. It remains to be seen whether a gene manipulated in this way would faithfully respond to its normal control parameters during growth, development or external stimuli.

IX Conclusion

In this review we have described the principles and scope of current techniques for manipulating DNA. Considerable advances have been made in gene cloning and the field is still expanding; the active researcher must carefully screen new publications for novel and improved methods.

The manipulation of genetic material has become widespread not only because it provides a quick and relatively easy answer to many biological questions, but also because of the ever increasing commercial availability of the enzymes, reagents and oligonucleotides necessary for even the most elementary manipulations.

We would like here to suggest a few brief guidelines for the design and execution of experiments. Many experiments run into unexpected difficulties not for want of technical information or expertise but rather from lack of planning. First, we would like to stress the versatility of unique restriction sites in any construction; in cases where more than one site exists for an enzyme, the use of that enzyme is seriously limited. It is often worthwhile to change an unwanted site to one of a different specificity by site conversion (section III) or linker insertion (section IV). Rarely is it useful to irreversibly destroy the recognition site for a restriction enzyme. By altering the sequence at one site, another site can be made unique.

We must additionally emphasise the unique characteristics of the single-stranded phage vectors. Quite apart from their utility in mutagenesis experiments (section VIII), their use as a means to generate rapid sequence information should not be overlooked. The ready availability of sequence data is perhaps the single most important aspect of DNA engineering, and analysis of a DNA sequence in advance gives the scientist an enormous advantage in experimental design.

Lastly, rapid improvements in the synthesis of artificial oligonucleotides have provided the DNA engineer with one of his most useful tools. They permit the joining and subsequent separation of diverse DNA fragments. In addition, the potential of defined sequence oligonucleotides to introduce specific alteration into a DNA sequence totally devoid of restriction sites cannot be underestimated.

The application of the methods and approaches described here, and future technical improvements and extensions, will continue to produce major progress in molecular biology and its commercial applications.

Acknowledgments

We would like to thank A. Brown, P. Tolstoshev, D. Gaffney and M. P. Kieny for critical reading of the manuscript. We are also grateful to M. P. Kieny and M. Lecocq for help in preparing Tables and Figs., and P. Gabriel for typing the text.

During the preparation of this review R. Everett was working in the laboratory of Professor Pierre Chambon and was supported by a Fellowship from the Université Louis Pasteur, Strasbourg.

XI References

Anderson, R. A., Krakauer, T. and Camerini-Otero, R. D. (1982). *Proc. Natn. Acad. Sci. U.S.A.* **79**, 2748–2752.

Anderson, S. (1981). *Nucl. Acids Res.* **9**, 3015–3027.

Armstrong, K. and Bauer, W. R. (1982). *Nucl. Acids Res.* **10**, 993–1007.

Backman, K. (1980). *Gene* **11**, 169–171.

Bahl, C. P., Marians, K. J., Wu, R., Stawinsky, J. and Narang, S. (1976). *Gene* **1**, 81–92.

Bahl, C. P., Wu, R., Brousseau, R., Sood, A. K., Hsiung, H. M. and Narang, S. (1978). *Biochem. Biophys. Res. Comm.* **81**, 695–703.

Battula, N. and Loeb, L. A., (1976). *J. Biol. Chem.* **251**, 982–986.

Beard, P., Morrow, J. F. and Berg, P. (1973). *J. Virol.* **12**, 1303–1313.

Bedbrook, J. R. and Gerlach, W. L. (1980). *In* "Genetic Engineering: Principles and Methods" (J. K. Setlow and A. Hollaender, eds) Vol. 2, 1–19. Plenum Press, New York and London.

Bird, A. P. (1980). *Nucl. Acids Res.* **8**, 1499–1504.

Bishop, J. O. (1979). *J. Molec. Biol.* **128**, 545–559.

Bolivar, F., Rodriguez, R. L., Greene, P. J., Betlach, M. C., Heynecker, H. L. and Boyer, H. W. (1977). *Gene* **2**, 95–113.

Boyer, H. W., Chow, L. T., Dugaiczyk, A., Hedgpeth, J. and Goodman, H. M. (1973). *Nature New Biol.* **244**, 40–43.

Brooks, J. E. and Roberts, R. J. (1982). *Nucl. Acids Res.* **10**, 913–934.

Burckhardt, J., Weisemann, J. and Yuan, R. (1981). *J. Biol. Chem.* **256**, 4024–4032.

Buryanov, Y. I., Bogdarina, I. G. and Bayev, A. A. (1978). *FEBS Letters* **88**, 251–254.

Busby, S., Irani, M. and de Crombrugghe, B. (1982). *J. Molec. Biol.* **154**, 197–209.

Chang, S. and Cohen, S. N. (1977). *Proc. Natn. Acad. Sci. U.S.A.* **74**, 4811–4815.

Charnay, P., Perricaudet, M., Galibert, F. and Tiollais, P. (1978). *Nucl. Acids Res.* **5**, 4479–4494.

Ciampi, M. S., Melton, D. A. and Cortese, R. (1982). *Proc. Natn. Acad. Sci. U.S.A.* **79**, 1388–1392.

Collins, J. and Hohn, B. (1979). *Proc. Natn. Acad. Sci. U.S.A.* **75**, 4242–4246.

Conley, A. J., Knipe, D. M., Jones, P. C. and Roizman, B. (1981). *J. Virol.* **37**, 191–206.

Di Maio, D. and Nathans, D. (1980). *J. Molec. Biol.* **140**, 129–142.

Doehmer, J., Barinaga, M., Vale, W., Rosenfeld, M. G., Verma, I. M. and Evans, R. M. (1982). *Proc. Natn. Acad. Sci.* **79**, 2268–2272.

Doel, M. T., Eaton, M. Cook, E. A., Lewis, H., Patel, T. and Carey, N. H. (1980). *Nucl. Acids Res.* **8**, 4574–4592.

Donoghue, D. J. and Hunter, T. (1982). *Nucl. Acids Res.* **10**, 2549–2564.

Dugaiczyk, A., Boyer, H. W. and Goodman, H. M. (1975). *J. Molec. Biol.* **96**, 171–184.

Engler, M. J. and Richardson, C. C. (1982). *In* "The Enzymes" (P. D. Boyer, ed.) Vol. XV, 3–29. Academic Press, New York.

Everett, R. D. and Chambon, P. (1982). *The E.M.B.O. J.* **1**, 433–437.

Ferretti, L. and Sgaramella, V. (1981a). *Nucl. Acids Res.* **9**, 85–93.

Ferretti, L. and Sgaramella, V. (1981b). *Nucl. Acids Res.* **9**, 3695–3705.

Flavell, R. A., Sabo, D. L., Bandle, E. F. and Weissmann, C. (1974). *J. Molec. Biol.* **89**, 255–272.

Fraser, M. J. (1980). *In* "Methods in Enzymology" (L. Grossman and K. Moldave, eds) Vol. 65, 255–263. Academic Press, London.

Gardner, R. C., Howarth, A. J., Messing, J. and Shepard, R. J. (1982). *DNA* **1**, 109–115.

Geier, G. E. and Modrich, P. (1979). *J. Biol. Chem.* **254**, 1408–1413.

George, J., Blakesley, R. W. and Chirikjian, J. G. (1980). *J. Biol. Chem.* **255**, 6521–6524.

George, J. and Chirikjian, J. G. (1982). *Proc. Natn. Acad. Sci. U.S.A.* **79**, 2432–2436.

Gillam, S., Astell, C. R. and Smith, M. (1980). *Gene* **12**, 129–137.

Giza, P. E., Schmit, D. M. and Murr, B. L. (1981). *Gene* **15**, 331–342.

Graessmann, M. and Graessmann, A. (1976). *Proc. Natn. Acad. Sci. U.S.A.* **73**, 366–370.

Gray, H. B., Ostrander, D. A., Hodnett, J. L., Legerski, R. J. and Robberson, D. L. (1975). *Nucl. Acids Res.* **2**, 1459–1493.

Gray, P. W., Leung, D. W., Pennica, D., Yelyerton, E., Najarian, R., Simonsen, C. C., Derynck, R., Sherwood, P. J., Wallace, D. M., Berger, S. L. Levinson, A. D. and Goeddel, D. V. (1982). *Nature, Lond.* **295**, 503–508.

Green, C. and Tibbetts, C. (1980). *Proc. Natn. Acad. Sci. U.S.A.* **77**, 2455–2459.

Greene, P. J., Poonian, M. S., Nussbaum, A. L., Tobias, L., Garfin, D. E., Boyer, H. W. and Goodman, H. M. (1975). *J. Molec. Biol.* **99**, 237–261.

Grippo, P., Daccararino, M., Parisie, E. and Scarano, E. (1968). *J. Molec. Biol.* **36**, 195–208.

Gronenborn, B. and Messing, J. (1978). *Nature, Lond.* **272**, 375–377.

Grosveld, G. C., Rosenthal, A. and Flavell, R. A. (1982). *Nucl. Acids Res.* **10**, 4951–4971.

Haberman, A., Heywood, J. and Meselson, M. (1972). *Proc. Natn. Acad. Sci. U.S.A.* **69**, 3138–3141.

Hartley, J. L. and Gregori, T. J. (1981). *Gene* **13**, 347–353.

Hattman, S. (1977). *J. Bacteriol.* **129**, 1330–1334.
Hattmann, S. (1981). *In* "The Enzymes" (P. D. Boyer, ed.) Vol. XIV, 517–548. Academic Press, New York.
Heffron, F., McCarthy, B. J., Ohtsubo, H. and Ohtsubo, E. (1979). *Cell* **4**, 1153–1163.
Herman, G. E. (1981). Ph.D. Thesis, Duke University.
Heynecker, H. L., Shine, J., Goodman, H., Boyer, H. W., Rosenberg, J., Dickerson, R. E., Narang, S. A., Itakura, K., Lin, S. Y. and Riggs, A. (1976). *Nature, Lond.* **263**, 748–752.
Hoopes, B. C. and McClure, R. R. (1981). *Nucl. Acids Res.* **9**, 5493–5504.
Humayan, M. S. and Chambers, R. W. (1979). *Nature, Lond.* **278**, 524–529.
Hsu, M. T. and Berg, P. (1978). *Biochemistry* **17**, 131–138.
Ish-Horowicz, D. and Burke, J. F. (1980). *Nucl. Acids Res.* **9**, 13.
Jay, G., Khoury, G., Seth, A. K. and Jay, E. (1981). *Proc. Natn. Acad. Sci. U.S.A.* **78**, 5543–5548.
Kai, K., Tsuruo, T. and Hayatsu, H. (1974). *Nucl. Acids Res.* **1**, 884–899.
Kalderon, D., Oostra, B. A., Ely, B. K. and Smith, A. E. (1982). *Nucl. Acids Res.* (in press).
Kan, N. C., Lautenberger, J. A., Edgell, M. H. and Hutchinson, C. A. (1979). *J. Molec. Biol.* **130**, 191–207.
Kapoor, Q. S. and Chinnadurai, G. (1981). *Proc. Natn. Acad. Sci. U.S.A.* **78**, 2184–2188.
Kemp, D. J., Cory, S. and Adams, J. M. (1979). *Proc. Natn. Acad. Sci. U.S.A.* **76**, 4627–4631.
Klenow, H. and Henningsen, I. (1970). *Proc. Natn. Acad. Sci. U.S.A.* **65**, 168–175.
Kornberg, A. (1980). "DNA Replication". Freeman, San Francisco.
Kramer, W., Schughart, K. and Fritz, H. J. (1982). *Nucl. Acids Res.* **10**, 6475–6485.
Kunkel, T. A. and Loeb, L. A. (1979). *J. Biol. Chem.* **254**, 5718–5725.
Kurtz, D. T. and Nicodemus, C. F. (1981). *Gene* **13**, 145–152.
Lacks, S. and Greenberg, B. (1977). *J. Molec. Biol.* **114**, 153–168.
Laskowski, M. (1980). *In* "Methods in Enzymology" (L. Grossman and K. Moldave, eds) Vol. 65, 263–276. Academic Press, London.
Lathe, R., Balland, A., Kohli, V. and Lecocq, J. P. 1982). *Gene* **20**, 185–193.
Lehman, R. (1981a). *In* "The Enzymes" (P. D. Boyer, ed.) Vol. XIV, 15–37. Academic Press, New York.
Lehman, R. (1981b). *In* "The Enzymes" (P. D. Boyer, ed.) Vol. XIV, 51–65. Academic Press, New York.
Lilley, D. M. (1980). *Proc. Natn. Acad. Sci. U.S.A.* **77**, 6468–6472.
Lilley, D. M. (1981). *Nucl. Acids Res.* **9**, 1271–1289.
Malcolm, A. D. B. (1981). *In* "Genetic Engineering" (R. Williamson, ed.) Vol. 2, 129–173. Academic Press, London.
Marinus, M. G. (1973). *Molec. Gen. Genet.* **127**, 47–55.
May, M. S. and Hattman, S. (1975). *J. Bacteriol.* **123**, 768–770.
Malyguine, E., Vannier, P. and Yot, P. (1980). *Gene* **8**, 163–177.
Maniatis, T., Hardison, R. C., Lacy, E., Lauer, J., O'Connell, C., Quon, D., Sim, G. K. and Efstradiadis, A. (1978). *Cell* **15**, 687–701.
McClelland, M. (1981a). *Nucl. Acids Res.* **9**, 5859–5866.
McClelland, M. (1981b). *Nucl. Acids Res.* **9**, 6795–6804.
McEntee, K., Weinstock, G. M. and Lehman, I. R. (1979). *Proc. Natn. Acad. Sci. U.S.A.* **76**, 2615–2619.
McKnight, S. L. and Kingsbury, R. (1982). *Science* **217**, 316–324.

Messing, J., Crea, R. and Seeburg, P. H. (1981). *Nucl. Acids Res.* **9**, 309–321.

Messing, J. and Vieira, G. (1982). *Gene* **19** (in press).

Miller, J. H. (1972). Experiments in Molecular Genetics. Cold Spring Harbor Laboratory Press, Cold Spring Harbor, N.Y.

Miyada, C. G., Soberon, X., Itakura, K. and Wilcox, G. (1982). *Gene* **17**, 167–177.

Modrich, P. (1979). *Quart. Rev. Biophys.* **12**, 315–369.

Modrich, P. and Lehman, I. R. (1973). *J. Biol. Chem.* **248**, 7502–7511.

Muller, W., Weber, H., Meyer, F. and Weissmann, C. (1978). *J. Mol. Biol.* **124**, 343–358.

Mulligan, R. C., Howard, B. H. and Berg, P. (1979). *Nature, Lond.* **277**, 108–114.

Murray, N. E., Bruce, S. and Murray, K. (1979). *J. Molec. Biol.* **132**, 493–505.

Murray, N. E. and Murray, K. (1974). *Nature, Lond.* **251**. 476–481.

Myers, R. M. and Tijan, R. (1980). *Proc. Natn. Acad. Sci. U.S.A.* **77**, 6491–6495.

Nilsson, S. V. and Magnusson, G. (1982). *Nucl. Acids Res.* **10**, 1425–1437.

Norris, K. E., Iserentant, D., Contreras, R. and Fiers, W. (1979). *Gene* **7**, 355–362.

Nosikov, V. V., Braga, E. A., Karlishev, A. V., Zhuze, A. L. and Polyanovsky, O. L. (1976). *Nucl. Acids Res.* **3**, 2293–2301.

Olivera, B. M. and Lehman, I. R. (1967). *Proc. Natn. Acad. Sci. U.S.A.* **57**, 1700–1704.

Pagano, J. S. (1970). *Prog. Med. Virol.* **12**, 1–48.

Panasenko, S. M., Alazard, R. J. and Lehman, I. R. (1978). *J. Biol. Chem.* **253**, 4590–4592.

Panasenko, S. M., Cameron, J., Davis, R. W. and Lehman, I. R. (1977). *Science* **196**, 188–189.

Panayotatos, N. and Truong, K. (1981). *Nucl. Acids Res.* **9**, 5679–5688.

Polinsky, B., Greene, B., Garfin, D. E., McCarty, B. J., Goodman, H. M. and Boyer, H. W. (1975). *Proc. Natn. Acad. Sci. U.S.A.* **72**, 3310–3314.

Radman, M., Wagner, R. E., Glickman, W. and Meselson, M. (1980). *In* "Progress in Environmental Mutagenesis" (M. Alacevic, ed.). Developments in Toxicology and Environmental Sciences, Vol. 7, 121–130, Elsevier, Amsterdam.

Roberts, R. J. (1981). *Nucl. Acids Res.* **9**, R75–R96.

Rosner, A., Bastos, R. N. and Aviv, H. (1978). *Molec. Biol. Rep.* **4**, 253–256.

Rothstein, R. J., Lau, L. F., Bahl, C. P., Narang, S. A. and Wu, R. (1979). *In* "Methods in Enzymology" (R. Wu, ed.) Vol. 68, 98–109. Academic Press, London.

Ruvkun, G. B. and Ausubel, F. M. (1981). *Nature, Lond.* **289**, 85–88.

Sandri-Goldin, R. M., Levine, M. and Glorioso, J. G. (1981). *J. Virol.* **38**, 41–49.

Sanger, F., Coulson, A. R., Barrell, B. G., Smith, A. J. H. and Roe, B. A. (1980). *J. Molec. Biol.* **143**, 161–178.

Scangos, G. A. and Ruddle, F. H. (1981). *Gene* **14**, 1–10.

Scangos, G. A., Huttner, K. M., Juricek, D. K. and Ruddle, F. H. (1981). *Molec. Cell Biol.* **1**, 111–120.

Schaffner, W. (1980). *Proc. Natn. Acad. Sci. U.S.A.* **77**, 2163–2167.

Setlow, P., Brutlag, D. and Kornberg, A. (1972). *J. Biol. Chem.* **247**, 224–231.

Scherer, S. and Davis, R. W. (1979). *Proc. Natn. Acad. Sci. U.S.A.* **76**, 4951–4955.

Sgaramella, V., Van de Sande, J. H. and Khorana, H. G. (1970). *Proc. Natn. Acad. Sci. U.S.A.* **67**, 1468–1475.
Shenk, T. E., Carbon, J. and Berg, P. (1976). *J. Virol.* **18**, 664–671.
Shepard, H. M., Yelverton, E. and Goeddel, D. V. (1982). *DNA* **1**, 125–131.
Shibata, T., Dasgupta, C., Cunningham, R. P. and Radding, C. M. (1979). *Proc. Natn. Acad. Sci. U.S.A.* **76**, 1638–1642.
Shishido, K. and Ando, T. (1981). *Biochim. Biophys. Acta* **656**, 123–127.
Shore, D., Langowski, J. and Baldwin, R. L. (1981). *Proc. Natn. Acad. Sci. U.S.A.* **78**, 4833–4837.
Shortle, D., DiMaio, D. and Nathans, D. (1981). *Ann. Rev. Genet.* **15**, 265–294.
Shortle, D. and Nathans, D. (1978). *Proc. Natn. Acad. Sci. U.S.A.* **75**, 2170–2174.
Shortle, D. and Nathans, D. (1979). *J. Mol. Biol.* **131**, 801–817.
Shortle, D., Grisafi, P. Benkovic, S. J. and Botstein, D. (1982). *Proc. Natn. Acad. Sci. U.S.A.* **79**, 1588–1592.
Shortle, D., Koshland, D., Weinstock, G. M. and Botstein, D. (1980). *Proc. Natn. Acad. Sci. U.S.A.* **77**, 5375–5379.
Solnick, D. (1981). *Nature, Lond.* **291**, 508–510.
Smith, M. and Gillam, S. (1981). *In* "Genetic Engineering: Principles and Methods" (J. K. Setlow and A. Hollaender, eds) Vol. 3, 1–32. Plenum Press, New York.
Streeck, R. (1980). *Gene* **12**, 267–275.
Sugino, A., Goodman, H. M., Heynecker, H. L., Shine, J., Boyer, H. W. and Cozzarelli, N. R. (1977). *J. Biol. Chem.* **252**, 3987–3994.
Taboni, C. J., Hoffman, J. W., Goff, S. P. and Weinberg, R. A. (1982). *Molec. Cell. Biol.* **2**, 426–436.
Traboni, C., Ciliberto, G. and Cortese, R. (1982). *The E.M.B.O. J.* **1**, 415–420.
Tye, B. K., Chien, J., Lehman, I. R., Duncan, B. K. and Warner, H. R. (1978). *Proc. Natn. Acad. Sci. U.S.A.* **75**, 233–237.
Tyndall, C., La Mantia, G., Thaker, C. M., Favaloro, J. and Kamen, R. (1981) *Nucl. Acids Res.* **9**, 6231–6250.
Ullrich, A., Shine, J., Chirgwin, J., Pictet, R., Tischer, E., Rutter, W. J. and Goodman, H. (1977). *Science* **196**, 1313–1319.
Van Wezenbeek, P. M. G. F., Hulsebos, T. J. M. and Schoenmakers, J. G. G. (1980). *Gene* **11**, 129–148.
Vardimon, L., Kressman, A., Cedar, H., Maechler, M. and Doerfler, W. (1982). *Proc. Natn. Acad. Sci. U.S.A.* **79**, 1073–1077.
Venegas, A., Mottes, M., Vasquez, C. and Vicuna, R. (1981). *FEBS Letters* **130**, 272–274.
Verma, I. M. (1981). *In* "The Enzymes" (P. D. Boyer, ed.) Vol. XIV, 87–103. Academix Press, New York.
Vogt, V. M. (1973). *Eur. J. Biochem.* **33**, 192–200.
Vogt, V. M. (1980). *In* "Methods in Enzymology" (L. Grossman and K. Moldave, eds) Vol. 65, 248–255. Academic Press, London.
Volker, T. A. and Showe, M. K. (1980). *Molec. Gen. Genet.* **177**, 447–452.
Waechter, D. R. and Baserga, R. (1982). *Proc. Natn. Acad. Sci. U.S.A.* **79**, 1106–1110.
Wallace, R. B., Johnson, P. F., Tanaka, S., Schold, M., Itakura, K. and Abelson, J. (1980). *Science* **209**, 1396–1400.
Wallace, R. B., Schold, M., Johnson, J. M., Dembek, P. and Itakura, K. (1981). *Nucl. Acids Res.* **9**, 3647–3656.
Wartell, R. M. and Reznikoff, W. S. (1980). *Gene* **9**, 307–319.
Wasylyk, B., Derbyshire, R., Guy, A., Molko, D., Roget, A., Teoule, R. and Chambon, P. (1980). *Proc. Natn. Acad. Sci. U.S.A.* **77**, 7024–7028.

Watabe, H. O., Shibata, T. and Ando, T. (1981). *J. Biochem.* **90**, 1623–1632.
Watanabe, S. M. and Goodman, M. F. (1981). *Proc. Natn. Acad. Sci. U.S.A.* **78**, 2864–2868.
Weber, H., Dierks, P., Meyer, F., van Ooyen, A., Dobkin, C., Abrescia, P., Kappeler, M., Meyback, B., Zeltner, A., Mullen, E. E. and Weissmann, C. (1981). *In* "Developmental Biology Using Purified Genes" (D. Brown and C. R. Fox, eds) Vol. 23, ICN-UCLA Symposium on Molecular and Cellular Biology, Academic Press, New York.
Weiher, H. and Schaller, H. (1982). *Proc. Natn. Acad. Sci. U.S.A.* **79**, 1408–1412.
Weiss, B. and Richardson, C. C. (1968). *J. Biol. Chem.* **243**, 4556–4563.
Weissmann, C., Nagata, S., Taniguchi, T., Weber, H. and Meyer, F. (1979). *In* "Genetic Engineering: Principles and Methods" (J. K. Setlow and A. Hollaender, eds) Vol. II, 73–92. Plenum Press, New York and London.
Wells, R. D., Klein, R. D. and Singleton, C. K. (1981). *In* "The Enzymes" (P. D. Boyer, ed.) Vol. XIV, 157–191. Academic Press, New York.
Wigler, M., Silverstein, S., Lee, L. S., Pellicer, A., Cheng, T. and Axel, R. (1977). *Cell* **11**, 223–232.
Wilson, G. G. and Murray, N. E. (1979). *J. Molec. Biol.* **132**, 471–491.
Woodbury, C. P., Hagenbuchle, O. and von Hippel, P. H. (1980). *J. Biol. Chem.* **255**, 11534–11546.
Wu, R., Bahl, C. P. and Narang, S. A. (1978). *Prog. Nucl. Acid Res. Molec. Biol.* **21**, 101–141.
Yuan, R. (1981). *Ann. Rev. Biochem.* **50**, 285–315.
Zakour, R. A. and Loeb, L. A. (1982). *Nature, Lond.* **295**, 708–710.
Zimmerman, F. K. (1977). *Mutat. Res.* **79**, 127–148.

Recombinant DNA technology: application to the characterisation and expression of polypeptide hormones

R. K. CRAIG and L. HALL

Courtauld Institute of Biochemistry, The Middlesex Hospital Medical School, Mortimer Street, London W1P 7PN, UK

I Introduction

The evolution of higher organisms, each comprised of tissues with different and specialised functions, required the parallel development of an integrated intercellular communication system. In animals, such a system has developed based on chemical signals produced by a complex endocrine system. The chemical signals, or hormones, modulate metabolic, developmental and behavioural responses by their action on target cells in different tissues. This communication system is itself regulated by a complex series of poorly understood events initiated in the brain, resulting in the release of neurotransmitter or neuromodulator molecules. These in turn act on certain endocrine glands culminating in the release of hormones into the circulation. An understanding of the structure and physiological function of peptide and steroid hormones, the factors which regulate their synthesis and secretion, and the mechanisms by which they modulate patterns of gene expression and rates of metabolic events in different target tissues, will ultimately provide insight into the complex communication systems which determine developmental and behavioural patterns in animals.

Previous chapters in this series have covered in depth the construction of cDNA and genomic DNA libraries (see Williams, 1981; Dahl *et al.*, 1981), and the construction and application of a variety of prokaryotic (Brammar, 1982; Thompson, 1982) and eukaryotic host/vector systems (Rigby, 1982).

The intention of this chapter is to consider the manner in which three related technologies, gene cloning, rapid DNA sequencing and the chemical synthesis of DNA, have widened our understanding of the structure and expression of polypeptide hormone genes in vertebrates; have led to the prospect of the synthesis of theoretically unlimited amounts of biologically active peptides of clinical, agricultural, and consequently economic importance; and have resulted in some instances, in the identification of cryptic peptide sequences, some of which have been shown to have biological activity. However, we do not intend to present an encyclopedic account of the applications of recombinant DNA technology to the study of peptide hormones, or for that matter make more than passing comments on the physiological role of the peptides considered.

II Historical perspectives

A The peptide hormones

Pioneering work, particularly in the 1960s and 1970s, resulted in the purification and structural characterisation of a number of large and small peptide hormones, the identification of their site(s) of synthesis, and determination of their predominant physiological roles. Much of this work relied on painstaking bioassay, radioimmunoassay and chromatographic procedures, though the identification using highly sensitive bioassay procedures of biologically active peptides at low tissue concentration ($\leqslant \mu$g/g tissue) resulted in the development of new fluorometric, chromatographic and chemical screening techniques. This permitted the isolation of novel peptides in sufficient amounts and purity to permit subsequent amino-acid sequence determination (see Stein, 1978; Tatemoto and Mutt, 1981). Prior knowledge of protein sequence, the site of synthesis *in vivo*, and the availability of antibodies raised against purified or synthetic peptides, have proved invaluable to the molecular biologist in the design and execution of his work.

Recently, there has been an increasing awareness that polypeptide hormones, growth factors from the serum, neuroregulators and neurotransmitters from the central nervous system may be grouped into families on the basis of structural and functional relatedness (see Blundell and Humbel, 1980). Since size, subunit composition, secondary structure, amino-acid sequence and the presence or otherwise of various prosthetic groups essential for bioactivity must all be considered when developing a cloning strategy, particularly where the expression of biologically active peptides is intended, we have divided the different families of peptide hormones broadly (with apologies to the purists) into four groups, based on structural considerations alone.

1 Large multi-chain glycoprotein hormones

The gonadotrophins — pituitary luteinising hormone (LH) and follicle stimulating hormone (FSH), thyroid stimulating hormone (TSH), and placental chorionic gonadotrophin (CG). Each is composed of two dissimilar non-covalently associated subunits designated α and β. Each subunit is glycosylated and maintained in a dimeric form by intrachain disulphide bonds (for review see Pierce and Parsons, 1981). Amino-acid sequence analysis shows that within the species studied, the α-subunit is common to all four hormones (human CG — 92 AAs), whereas the β-subunits vary in size (human CG — 147 AAs, human FSH — 118 AAs), contain sequences unique to each hormone, and confer biological specificity. Evidence suggests that the carbohydrate moieties play a functional role in hormonal activity (Saxena and Rathnam, 1976; Bahl, 1977).

2 *Large single-chain hormones*

The placental hormone, chorionic somatomammotrophin (placental lactogen), and the pituitary hormones prolactin and growth hormone. All are comprised of a single polypeptide chain (in human 191—198 AAs) of known amino-acid sequence, show homology and to a lesser extent overlapping biological activity (see Niall *et al.*, 1971). We would also include parathyroid hormone (in human 84 AAs — Habener *et al.*, 1978) and some growth factors (Rinderknecht and Humbel, 1978; Svoboda *et al.*, 1980) within this group.

3 *Small multi-chain hormones*

The structurally related ovarian and pancreatic hormones, insulin and relaxin. Each consists of two non-identical peptide chains aligned and linked by disulphide bonds (see James *et al.*, 1977).

4 *Small single-chain hormones*

A vast array of peptides ranging in size from the opiate pentapeptides Met- and Leu-enkephalin (Hughes *et al.*, 1975) to for instance calcitonin, which consists of 32 AAs and requires a carboxyl terminal amide structure for biological activity (see MacIntyre *et al.*, 1980), a feature common to many small biologically active peptides (see Tatemoto and Mutt, 1981). Many small peptide hormones are found in a variety of tissues including the brain, and therefore are considered to be potential neurotransmitters or neuromodulator molecules (see Krieger *et al.*, 1980; Herbert, 1981).

B Precursors and processing

Proteins destined for secretion are synthesised on polyribosomes associated with membranes of the endoplasmic reticulum. The nascent peptides pass through the membrane into the lumen of the endoplasmic reticulum, and once sequestered, remain encapsulated within a membranous environment as they traverse the different intracellular organelles which comprise the secretory pathway, finally to be discharged from the cell, by fusion of secretory granules with the plasma membrane (see Palade, 1975; also Fig. 1).

Studies on the mechanisms involved in the secretion of proteins soon demonstrated that those destined for export were often synthesised as higher molecular weight precursor forms. Some of these precursors, classically the zymogens (digestive enzymes, blood clotting factors), were processed by proteolytic enzymes after secretion, whilst others including peptide hormone precursors were generally cleaved and otherwise modified within the cell before secretion occurred (for review see Steiner *et al.*, 1980).

In some instances the role of the additional peptide region(s) in hormone precursors was self evident. Purification and amino-acid sequence analysis of proinsulin from the β-cells of the pancreas demonstrated that the A and B chains of insulin are synthesised as a single polypeptide chain linked by an additional C peptide region. The role of the C-peptide is to ensure the correct orientation of disulphide bridges of the product molecule insulin. Once these bridges had been formed, the C-peptide is removed by proteolytic cleavage by enzymes in the secretory granules with specificities similar to trypsin and carboxypeptidase B, at sites specified by paired basic (Arg or Lys) amino-acid residues (see Steiner *et al.*, 1974, also Fig. 1). Isolation and characterisation of proparathyroid hormone from the parathyroid gland produced a similar picture of a prohormone intermediate, cleaved to an active form at a cleavage site determined by basic residues, in this instance by proteolytic enzymes in the Golgi vesicles (Habener and Potts, 1978a, b). No biological activity has been ascribed to the C-peptide of insulin or the hexapeptide "pro" sequence of parathyroid hormone.

The search for intermediates in the biosynthesis of other peptide hormones, in tissues and cell lines, proved equally fascinating. High molecular weight forms of many small peptide hormones including glucagon, somatostatin, gastrin, arginine vasopressin, oxytocin, the enkephalins, ACTH and β-lipotrophin have all been reported. In some instances it proved possible to demonstrate that several small peptides of known biological activity could be generated by the proteolytic cleavage of a single precursor or polyprotein. Thus the largest precursor form of $ACTH_{1-39}$ (proopiocortin — estimated molecular weight 30 000) synthesised by a mouse pituitary cell line proved to be processed via glycosylated intermediates to yield ACTH, β-lipotropin, β-endorphin (a potent opioid peptide), and other peptide fragments not fully characterised due to scarcity of material (see Mains *et al.*, 1977; Mains and Eipper, 1980). Moreover, evidence points to different processing pathways of the proopiocortin precursor dependent on the site of synthesis within the pituitary (Mains and Eipper, 1980) and whether expressed in adult or fetal pituitary tissues (Silman *et al.*, 1978). Recent studies point to the presence of an arginyl protease activity associated with a secretory granule fraction of the rat pituitary neurointermediate lobe, capable of producing some, but not all, of the expected cleavage products of proopiocortin (Loh and Gainer, 1982). Similar prohormone-converting proteases capable of cleaving prosomatostatin

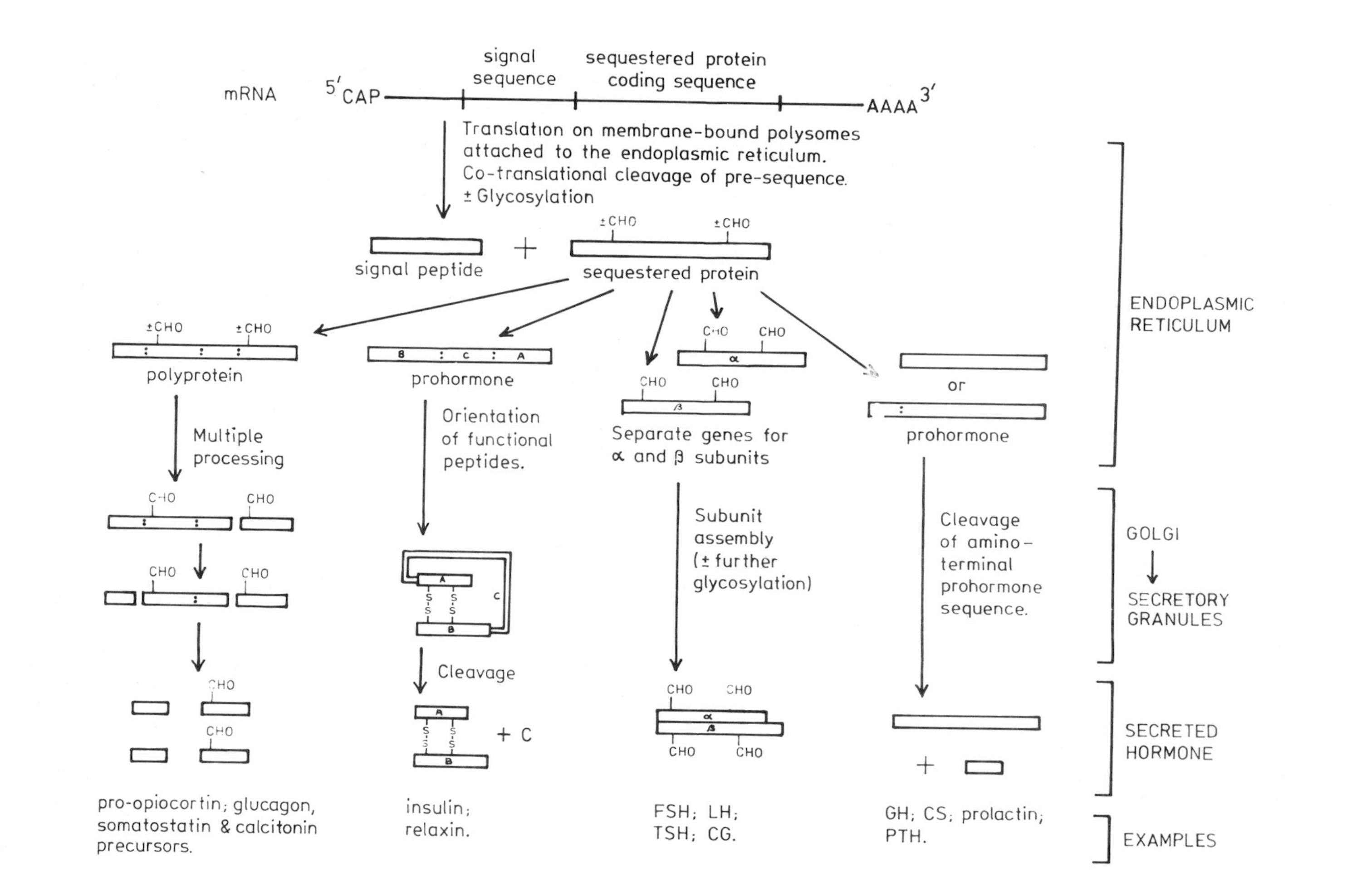
mRNA
5′ CAP
AAAA 3′
signal sequence
sequestered protein coding sequence
Translation on membrane-bound polysomes attached to the endoplasmic reticulum.
Co-translational cleavage of pre-sequence.
± Glycosylation
signal peptide
sequestered protein
±CHO
polyprotein
prohormone
Separate genes for α and β subunits
or
prohormone
Multiple processing
Orientation of functional peptides.
Subunit assembly (± further glycosylation)
Cleavage of amino-terminal prohormone sequence.
Cleavage
+ C
CHO
pro-opiocortin; glucagon, somatostatin & calcitonin precursors.
insulin; relaxin.
FSH; LH; TSH; CG.
GH; CS; prolactin; PTH.
ENDOPLASMIC RETICULUM
GOLGI
SECRETORY GRANULES
SECRETED HORMONE
EXAMPLES

(Fletcher *et al.*, 1980), proinsulin and proglucagon (Fletcher *et al.*, 1981) have also been identified.

Evidence based on pulse-labelling experiments and peptide analysis also point to common precursors of vasopressin, oxytocin and their respective neurophysin carrier proteins (Brownstein *et al.*, 1980), whilst a common precursor containing both Leu-enkephalin and Met-enkephalin has been identified in the chromaffin cells of the adrenal medulla (Kimura *et al.*, 1980; Rossier *et al.*, 1980). Recently a processing enzyme for the enkephalin precursors with carboxypeptidase activity has been identified in the adrenal medullary chromaffin granules (Hook *et al.*, 1982).

C Messenger RNA-directed cell-free protein synthesis

The development of cell-free protein synthesising systems based on wheatgerm extracts (Roberts and Paterson, 1973) or rabbit reticulocyte lysate (Pelham and Jackson, 1976), in combination with simple mRNA isolation procedures (Aviv and Leder, 1972), provided fresh impetus for studies concerned with intracellular mechanisms involved in the biosynthesis and secretion of peptide hormones.

The application of *in vitro* synthesising systems in combination with microprotein sequencing techniques soon demonstrated that, in common with most other secretory proteins, all large peptide hormones and prohormone precursors studied were synthesised with hydrophobic NH_2-terminal "signal" peptide extensions (see Blobel and Dobberstein, 1975; Chan *et al.*, 1976; Habener *et al.*, 1978). As a result of innumerable studies using cell-free systems in the presence and absence of microsomal membranes, there is now general agreement that interaction of the nascent signal peptide in combination with other protein factors (Walter and Blobel, 1981; Meyer *et al.*, 1982) with the membranes of the endoplasmic reticulum, is responsible for the segregation of the newly synthesised secretory proteins within the lumen of the endoplasmic reticulum. During this process the signal peptide is generally, though seemingly not invariably, removed, a co-translational event (see Steiner *et al.*, 1980; Lingappa and Blobel, 1980). In some instances core glycosylation of sequestered *in vitro* synthesised peptide hormone precursors at the level of the

---

Figure 1 Alternative routes for synthesis, processing and assembly of peptide hormones. Peptide hormones may be broadly divided into 4 different classes, on the basis of size, subunit composition and mechanism of synthesis as outlined in the text. Glycosylation (represented as CHO) may occur at various points within the secretory pathway, as may proteolytic processing at characteristic cleavage sites (represented as :).

endoplasmic reticulum has also been demonstrated; in particular, subunits of the glycoprotein hormones (Bielinska and Boime, 1979; Kourides *et al.*, 1979) and high molecular weight precursors of small peptide hormones (see Jacobs *et al.*, 1981).

Signal peptides and segregation mechanisms apart, the synthesis of peptide hormones using mRNA directed cell-free systems, in conjunction with definitive analysis of the primary translation products, has confirmed and extended our knowledge of the intracellular mechanisms involved in the synthesis of peptide hormones both large and small. Thus it has proved possible to provide substantiating evidence (see also Fig. 1) that: (1) the proopiocortin (Roberts and Herbert, 1977a, b; Nakanishi *et al.*, 1977) and the neurophysin precursors (Schmale *et al.*, 1979; Schmale and Richter, 1980) are polyproteins containing the sequence of two or more peptides of known biological function; (2) that numerous small peptide hormones are synthesised as high molecular weight precursor proteins containing other cryptic peptide sequence(s) of unknown biological significance, and so may yet prove to be polyproteins; and (3) that the subunits of the gonadotrophins are synthesised as precursors from separate mRNA species, not via a larger common precursor protein (Daniels-McQueen *et al.*, 1978; Vamvakopoulos and Kourides, 1979; Godine *et al.*, 1980) and are glycosylated to some extent at least, at the level of the endoplasmic reticulum.

III Identification and characterisation of peptide hormone mRNA sequences

A The source of mRNA

The more abundant the mRNA one wishes to identify and subsequently clone within a given mRNA population the better — rather an obvious maxim with which to initiate any overall cloning strategy. However a little forethought can eliminate much frustration at a later stage.

Peptide hormones are synthesised and secreted by highly specialised cell types within endocrine tissues. It is reasonable to assume that in a given endocrine tissue under physiological conditions which favour maximum expression of a particular peptide hormone gene, the relative abundance of the mRNA species in question may represent anything between 10% and 0.1% of the total cellular mRNA. Thus physiological conditions should be chosen which coincide with maximum synthesis of the peptide hormone and therefore inevitably maximum mRNA levels within the chosen tissue. For instance, mRNA species encoding the two subunits of human chorionic gonadotrophin represent a major species in first-trimester placental mRNA populations, but are of lesser abundance in the full-term placenta (Daniels-McQueen *et al.*, 1978). Conversely, human chorionic somatomammotrophin represents the major mRNA species in the full term, but is less abundant in first-trimester placental mRNA (Boime *et al.*, 1976).

Relative mRNA concentrations may also be manipulated *in vivo*; for example the relative levels of the mRNA species encoding pituitary luteinising and follicle-stimulating hormone rise 10-fold in ovariectomised rats (Godine *et al.*, 1980). Alternatively, enrichment may be obtained by tissue dissection, particularly if the cells synthesising the peptide of interest represent only a small proportion of the total tissue population. Using this approach an mRNA population enriched in rat preproinsulin mRNA relative to total pancreatic mRNA was isolated from rat islets of Langerhans by collagenase digestion (Chan *et al.*, 1976), whilst enrichment of bovine prepro-opiocortin mRNA was achieved by isolation from the intermediate lobe of the bovine pituitary where the mRNA is at a higher concentration than in the anterior lobe (Taii *et al.*, 1979).

When human mRNA sequences are required, or where under normal circumstances, the mRNA sequences are either of low abundance, or synthesised by a small subset of the cellular population, then tumour tissue, or transplantable tumours (see Vamvakopoulos and Kourides, 1979) known to synthesise the hormone(s) in question must be used as a source of mRNA. Human calcitonin is synthesised by a small proportion of thyroid cells, the C-cells, representing as little as 2% of the cell population, and not readily amenable to cell fractionation techniques. Consequently postoperative tissue obtained from patients with medullary carcinoma of the thyroid, where circulating calcitonin levels are elevated up to 1000-fold above normal, provides a rich source of C-cells and human preprocalcitonin mRNA (Allison *et al.*, 1981; see Fig. 2). Pituitary tumour tissues obtained from patients with acromegaly or hyperprolactinemia are also suitably enriched in growth hormone and prolactin mRNA respectively (Martial *et al.*, 1979), whilst human pheochromocytoma tissue has recently been used as a source of the enkephalin precursor mRNA (Comb *et al.*, 1982a). Cell lines derived from tumour tissue have been used surprisingly rarely as a source of peptide hormone mRNA, possibly a reflection of the problems involved in the establishment of stable cell-lines which continue to synthesise the hormone of interest in significant amounts. However, one rat pituitary cell line (GC) which synthesises elevated levels of growth hormone mRNA in response to administration of glucocorticoid has been used as a source of mRNA for cDNA cloning (Seeberg *et al.*, 1977a; Harpold *et al.*, 1978) whilst a mouse pituitary cell line (AtT-20/D-16v) has

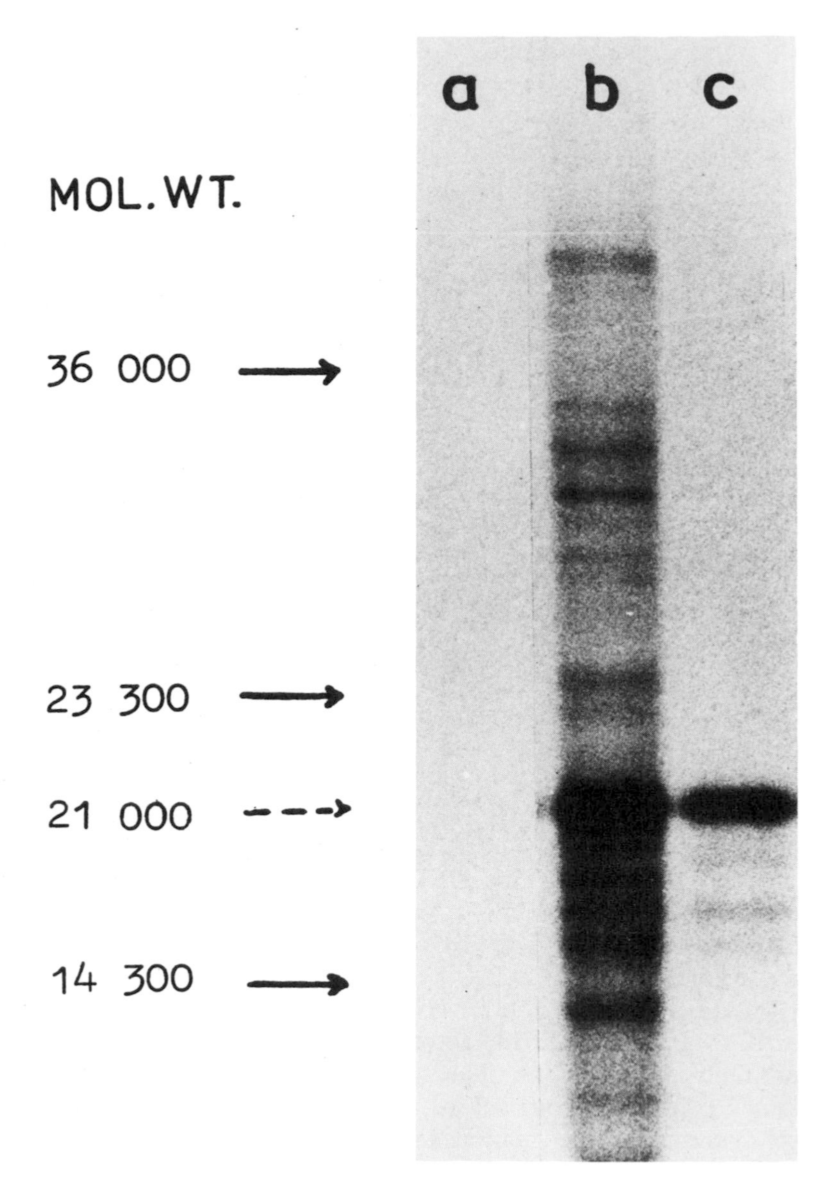

Figure 2 mRNA-directed cell-free protein synthesis of human calcitonin precursor polyprotein as judged by immunoprecipitation and SDS-polyacrylamide gel electrophoresis. Total poly(A)-containing RNA was isolated from frozen human thyroid medullary carcinoma tissue as described by Hall *et al.* (1979), translated in a wheat-germ cell-free protein-synthesizing system, and the resulting [^{35}S]-methionine-labelled proteins separated by SDS-polyacrylamide gell electrophoresis then visualized using fluorography. (a) no added mRNA; (b) human thyroid medullary carcinoma mRNA; (c) immunoprecipitation of proteins synthesized in (b) using an antiserum raised against synthetic human calcitonin. (reprinted from Allison *et al.*, 1981 — with permission of the Biochemical Journal).

been used extensively to study the biosynthesis and processing of proopiocortin (see Herbert *et al.*, 1980; Mains and Eipper, 1980) and as a source of preproopiocortin mRNA for cDNA cloning (Roberts *et al.*, 1979).

B Isolation of mRNA

Several considerations are important at this point, reflecting primarily the relative abundance of the peptide hormone mRNA within the tissue to be handled, the amount of tissue or cells in culture available, and the methods employed to identify a given mRNA species within a total or enriched mRNA population. The primary consideration is whether or not to isolate total cellular RNA, then isolate the poly(A)-containing RNA population, or alternatively, isolate membrane-bound polyribosomes (enriched in mRNA for secretory proteins) thus subjecting the tissue or cells to subcellular fractionation procedures, and thereby increasing the risk of degradation of the poly(A)-containing RNA population by endogenous and exogenous ribonucleases prior to deproteinization. Generally speaking, provided certain sensible precautions are taken — use sterile polyware, glassware, and buffers, wear gloves — degradation by exogenous ribonucleases is minimal. Where animal model systems or cell lines are under investigation, and consequently material is not at a premium, a preliminary fractionation in the presence of ribonuclease inhibitors (heparin, diethyl pyrocarbonate, etc) to give a membrane-bound polyribosomal fraction can provide a useful enrichment step (see Roberts and Herbert, 1977a; Schmale *et al.*, 1979). Whether time spent in the optimisation of subcellular fractionation procedures to give high yields of intact poly(A)-containing RNA is well spent, must be judged from experiment to experiment, but is probably only worthwhile where a relatively low abundance peptide hormone mRNA is to be cloned, and a highly enriched mRNA fraction is then required at a later stage for preliminary screening of the resulting recombinant plasmids (see Land *et al.*, 1982 — also section IV.C below). However, if the tissue is precious (scarce), then extract total nucleic acid immediately; do not risk subcellular fractionation procedures likely to release endogenous ribonucleases.

Two simple and related methods have been developed for the isolation of intact total nucleic acid from fresh or frozen tissue, or by direct lysis of cells in tissue culture. Each relies on the inhibition or inactivation of endogenous ribonucleases released during tissue extraction, one by rapid cell lysis and protein denaturation in 4 M guanidine thiocyanate (see Ullrich *et al.*, 1977; Chirgwin *et al.*, 1979), the other by cell lysis and protein denaturation by 2% (w/v) sodium dodecyl sulphate, in the presence of proteinase K (see Hall *et al.*,

1979), a broad spectrum proteolytic enzyme which shows maximum activity under such conditions (Hilz *et al.*, 1975). Both procedures then require the separation of total RNA from the bulk genomic DNA by equilibrium centrifugation through a 5.7 M caesium chloride cushion (Glisin *et al.*, 1974). In reality some DNA co-sediments with the RNA, and consequently a brief incubation at 4 °C with pancreatic deoxyribonuclease I is often advisable after careful inactivation of contaminating ribonucleases in the deoxyribonuclease preparations by treatment with iodoacetate as described by Zimmerman and Sandeen (1966). Commercial ribonuclease-free pancreatic deoxyribonuclease I preparations do not exist, whatever the labels claim! Unfortunately no single method of extraction provides a foolproof procedure for all tissues. Consequently, if the two methodologies outlined above proves disappointing, alternatives such as direct homogenisation into phenol (Liarakos *et al.*, 1973) or extraction in LiCl-urea (Auffray and Rougeon, 1980) should be evaluated.

The isolation of poly(A)-containing RNA from total cellular or polyribosomal RNA preparations using columns of oligo(dT)-cellulose (Aviv and Leder, 1972) or poly(U)-Sepharose (Palatnik *et al.*, 1979) is well established. Each matrix has it's disciples. We have routinely used oligo(dT)-cellulose, and find ribosomal RNA contamination minimal, provided that two successive cycles of binding are employed at 100 mM NaCl, with elution of the bound poly(A)-containing RNA in low salt at 60 °C (Craig *et al.*, 1976). Binding in 500 mM NaCl markedly increases the ribosomal RNA content of the final preparation. Such preparations may provide an "active" mRNA preparation as judged by cell-free protein synthesis, but give poor yields of cDNA and increase the chances of cloning ribosomal RNA sequences (see Williams, 1981). The use of SDS/proteinase K extraction and oligo(dT)-cellulose procedures as described above, should yield upward of 40 μg of intact biologically active poly(A)-containing RNA from 1 g wet weight of typical endocrine tissue, with the proviso that the tissue has been snap frozen in liquid N_2 and then stored at − 70 °C until required. Do not expect stunning results using human tissues which have been roasted in the pathology laboratory.

C Identification of mRNA

1 mRNA-directed protein synthesis

The criterion of an intact poly(A)-containing RNA generally resides in its ability to direct the synthesis of a wide molecular weight spectrum of ^{35}S-methionine labelled proteins, on addition to cell-free protein synthesising systems. Ideally, if a peptide hormone mRNA is present in abundance, then the primary translation product, as synthesised in the wheat-germ or rabbit reticulocyte cell-free systems, may be identified by antibody precipitation followed by SDS/polyacrylamide gel electrophoresis and fluorography (see Bonner and Laskey, 1974). The example (Fig. 2) shows the mRNA-directed synthesis in the wheat-germ cell-free system and subsequent immunoprecipitation using an antibody raised against synthetic human calcitonin (Girgis *et al.*, 1980) of a high molecular weight precursor(s) to human calcitonin (MW, 3500) using total poly(A)-containing RNA isolated from a medullary carcinoma of the thyroid.

Unfortunately, supposedly monospecific antibodies as judged by radioimmunoassay criteria may often prove non-specific or "sticky" when used in cell-free systems. Furthermore, antibodies raised against the mature secreted proteins may not necessarily recognise precursor forms, due to conformational differences in protein structure, or the absence of post-translational modifications. For instance, antiserum raised against purified human chorionic gonadotrophin reacts poorly with the non-glycosylated primary translation products of the individual subunits (Fiddes and Goodman, 1979), whilst antisera raised against native and denatured forms of the β-thyroid stimulating hormone subunit immunoprecipitated only microsomal membrane processed (presumed glycosylated) and primary translation products of βTSH-mRNA respectively (Giudice and Weintraub, 1979). Similarly antiserum raised against the β-MSH and γ-MSH immunoprecipitated the primary translation product of the preproopiocortin precursor protein yet antiserum raised against an NH_2-terminal fragment of ACTH failed to precipitate the same primary translation product (Silman and Holland, unpublished observations).

Although problems of product identification related to processing of primary translation products can to a limited extent be overcome by mRNA-directed protein synthesis in the presence of microsomal membranes, followed by analysis of the sequestered products, a better approach involves the injection of mRNA preparations into *Xenopus* oocytes. Oocytes have been shown to sequester, glycosylate, and secrete a number of mammalian secretory proteins as a result of microinjected mRNA (Colman and Morser, 1979; Lane *et al.*, 1980; Colman *et al.*, 1981). The oocyte has the additional advantage that small amounts of mRNA are translated with great efficiency over long periods. Thus, the oocyte has been used to demonstrate the biosynthesis and glycosylation of the α and β subunits of mouse thyroid stimulating hormone (Kourides *et al.*, 1979), and the bovine arginine vasopressin-neurophysin II precursor (Schmale and Richter,

1980). Interestingly, although the oocyte will glycosylate, it is unable to perform the proteolytic cleavage of sequestered or secreted forms of carp proinsulin (Rapoport, 1981), suggesting the absence of specialised proteolytic processing enzymes (excluding signal peptidases) within the oocyte secretory pathway. This result implies that these enzymes may prove, not surprisingly, to be cell or tissue specific, not obligatory components of the secretory pathway.

Fortunately, not all biologically active proteins require extensive cell or tissue-specific post-translational processing/modification to ensure that the secreted protein is active as judged by a bioassay. Thus the oocyte can be used as a highly efficient means of detecting the presence or otherwise of low abundance mRNA species which encode biologically active proteins, by the simple expedient of assaying the surrounding incubation medium, 24—48 h after mRNA injection, for the presence or otherwise of the required biological activity. This phenomenon was first described by Colman and Morser (1979), who showed that interferon activity could be detected in oocyte secretions after injection of total poly(A)-containing RNA from human interferon producing cells. The observation has proved to be of considerable importance, since it provides the means by which mRNA species encoding growth factors may be identified, permits these to be enriched for the required mRNA species by size fractionation, and provides the basis for the selection of recombinant plasmids containing cDNA sequences of interest (see IV.C.2). In addition to the identification of mRNA-directed interferon synthesis, oocyte microinjection procedures coupled with a bioassay have been used to demonstrate the presence of mRNA encoding granulocyte-macrophage colony stimulating factor activity (a glycoprotein), within a total poly(A)-containing RNA population isolated from a human T-lymphocyte cell-line (Lusis *et al.*, 1982).

2 *Hybridisation using synthetic oligodeoxynucleotides*

Suppose the mRNA species in question is only of moderate abundance, to the extent that time-consuming subcellular and size fractionation procedures are required to provide mRNA populations sufficiently enriched in the mRNA of interest to allow meaningful product analysis. Alternatively, suppose neither suitable antibody preparations, nor a sensitive bioassay are available. In such instances, provided that some amino-acid sequence is known, a different route may be taken to determine the presence, size and partial nucleotide sequence of a specific mRNA species within a poly(A)-containing total RNA population. This involves the use of synthetic oligodeoxynucleotides complementary to the mRNA sequence, as predicted from known amino-acid sequence, provided that degeneracy of the genetic code is minimal (see Itakura and Riggs (1980), for an account of the applications of chemical DNA synthesis to recombinant DNA studies).

The potential of synthetic oligodeoxynucleotide primers for the identification of peptide hormone mRNA species was first realised by the identification and partial sequence analysis of gastrin mRNA from a hog antral mucosa total mRNA preparation. In these experiments a dodecamer was synthesised complementary to the predicted mRNA sequence encoding a tetrapeptide known to be present in gastrin:

Gastrin tetrapeptide	Trp^{NH_2}	Met	Glu	Glu^{COOH}
			A	A
Predicted gastrin mRNA	5′-UGG	AUG	GAG	GAG-3′
Synthesised cDNA	3′-ACC	TAC	CTC	CTC-5′

The construction took into account the preferential occurrence of G or C in the third position of mammalian codons (see Grantham *et al.*, 1981), when considering the two potential codons for glutamic acid, but otherwise could not be affected by degeneracy of the genetic code, since only single codons are utilised for methionine and tryptophan. The synthetic oligodeoxynucleotide was then used to prime cDNA synthesis by reverse transcription, the resulting cDNA transcripts isolated, and the nucleotide sequence then determined (see Noyes *et al.*, 1979). Recently the potential of this procedure has been increased by the demonstration, using carefully defined experimental conditions, that G-dT or dG-U mismatches are in some instances permissable in short DNA-RNA hybrids. Allowing for these, an additional 9 amino acids — Phe, Tyr, His, Asn, Asp, Cys, Gln, Glu, Lys — are useful for predicting the sequence of oligodeoxynucleotide hybridisation probes. For instance the "deduced" cDNA for codons AAA and AAG (Lys) would be TTT, and the "deduced" cDNA for the codons UUU and UUC (Phe) would be AAG (see Agarwal *et al.*, 1981). Even in situations where application of the dG-U/G-dT mismatch rule fails to produce a single defined oligodeoxynucleotide probe, or alternatively proves unsuccessful since a single base mismatch can be the difference between success and failure, then provided that stringent hybridisation conditions are employed which reflect the base composition of the probe, a mixture of a number of possible oligodeoxynucleotides may be used successfully to identify the desired mRNA species (see Wallace *et al.*, 1981; Kakidani *et al.*, 1982).

Partial characterisation of the proenkephalin mRNA from the bovine adrenal medulla (Gubler *et al.*, 1981; see also Comb *et al.*,

1982a, b for a similar approach using human tissue) has provided a powerful demonstration of the use of synthetic probes to detect and partially characterise a relatively low abundance mRNA species (0.05—0.1% of poly(A)-containing RNA). In this instance the strategy was based on the G-U mismatching rule described above, using a synthetic probe complementary to an amino-acid sequence known to be present in one of the enkephalin-containing precursor peptides (see Stern *et al.*, 1981) and chosen because of its minimum possible variation in codon usage:

Peptide	NH_2	Trp	Trp	Met	Asp	Tyr	Gln	COOH
					U	U	A	
Predicted mRNA	5′	UGG	UGG	AUG	GAC	UAC	CAG	3′
Synthesised cDNA primer	3′	ACC	ACC	TAC	CTG	ATG	G	5′

Once synthesised, the decahexamer, ^{32}P-labelled at the 5′ end using T_4 polynucleotide kinase and [γ-^{32}P]ATP, was used to prime the synthesis of cDNA. This was then fractionated by gel electrophoresis, the major bands eluted, and their nucleotide sequence determined by the chemical cleavage method (see Maxam and Gilbert (1980), also section IV.D) in this instance confirming the specificity of the probe, since the sequence information obtained included a region of the mRNA encoding the Met-enkephalin pentapeptide. The ^{32}P-labelled cDNA synthesised in this manner, or the ^{32}P-labelled decahexamer alone, were also used to determined the size of the proenkaphalin mRNA using RNA blotting techniques (see Fig. 3).

Provided a source of total poly(A)-containing RNA is readily available, oligodeoxynucleotide primers may also be used directly to obtain nucleotide sequence using the chain termination procedures of Sanger *et al.* (1977), essentially as described for the sequencing of immunoglobulin mRNA species (Hamlyn *et al.*, 1978, 1981), and human fibroblast interferon (Houghton *et al.*, 1980).

The application of synthetic oligodeoxynucleotide hybridisation probes of predicted nucleotide sequence to the study of the proenkephalin mRNA provides a classic example of how a limited knowledge of amino-acid sequence may be used to identify the presence within a given RNA population of a previously unidentified mRNA species, its size and partial nucleotide sequence. In addition to the requirements outlined above, the ability to synthesise single and double-stranded oligodeoxynucleotides is of increasing importance for the selection of recombinant phage or plasmids from genomic and cDNA libraries, and the construction of recombinants where expression of the inserted sequence is intended (see Section V, also Harris, 1982).

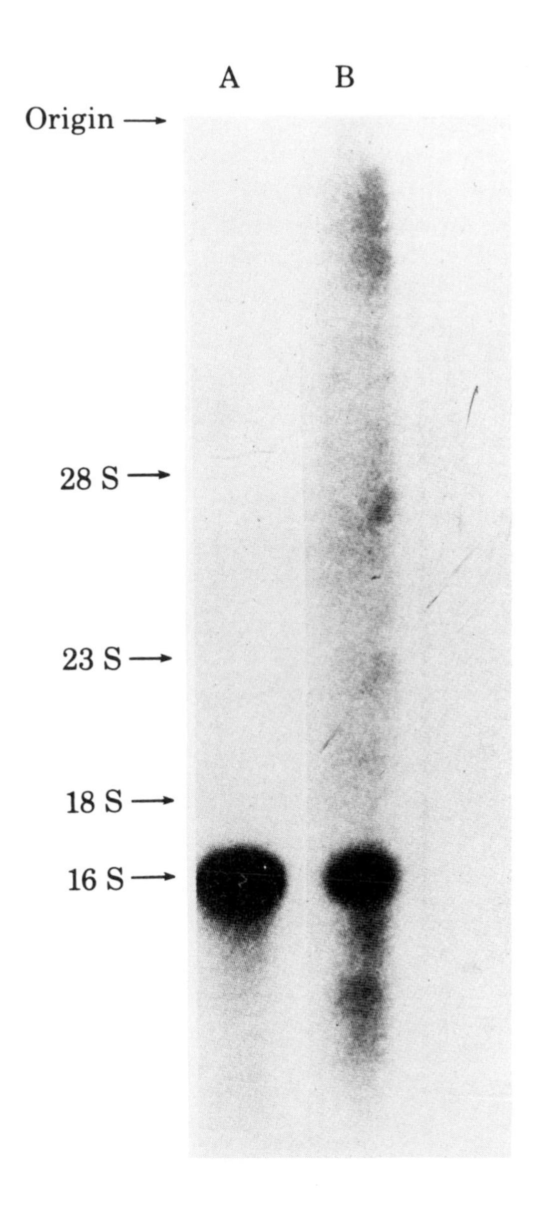

Figure 3 Hybridisation of ^{32}P-labelled decahexamer or cDNA to adrenal medullary poly(A)-containing RNA. Poly(A)-containing RNA was fractionated by agarose gel electrophoresis under denaturing conditions then transferred to nitrocellullose sheets. Adjacent lanes were then incubated under hybridising conditions with ^{32}P-labelled decahexamer or cDNA. Filters were washed, dried and autoradiographed at -70°C. Size markers were *Escherichia coli* rRNA and bovine rRNA, run in parallel lanes and visualized by staining with ethidium bromide prior to transfer. Lanes: A, decahexamer; B, cDNA. (reprinted from Gubler *et al.*, 1981 with permission of Dr U. Gubler).

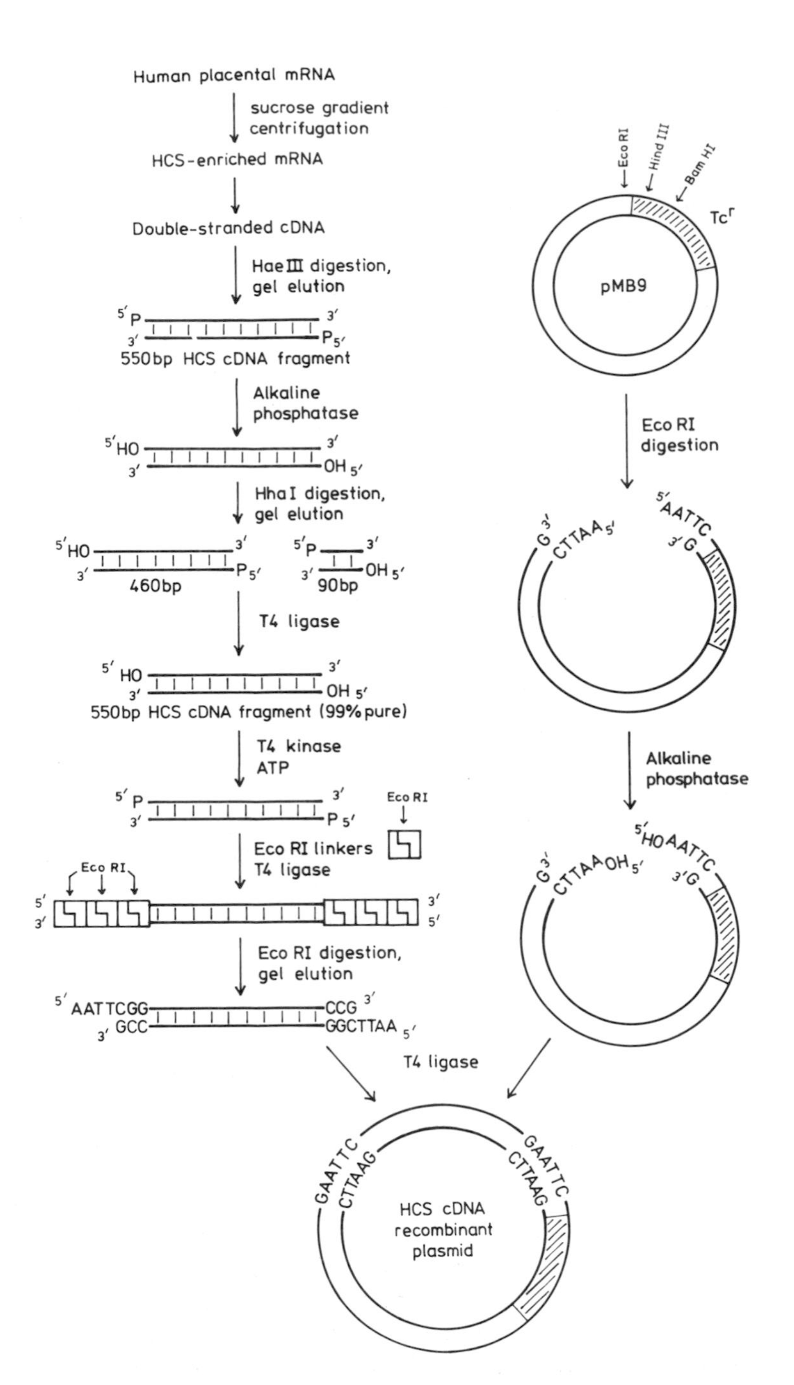

Figure 4 Construction of a recombinant plasmid containing a human chorionic somatomammotrophin cDNA sequence — a strategy for cloning defined cDNA sequences using chemical linkers (adapted from Seeberg *et al.*, 1977b).

IV Construction, identification and characterisation of cDNA plasmids

A Cloning defined cDNA sequences using linkers

Early experiments designed to construct cDNA clones containing peptide hormone sequences were limited by both physical and experimental constraints imposed due to conjectural hazards associated with the possibility that biologically active peptides might be expressed in bacteria, and that these in turn might have an adverse effect on their environment. One result of these constraints was the development of a cloning strategy using chemical linkers, outlined in Fig. 4, suitable for the cloning of abundant mRNA sequences. Although this strategy is not to be recommended in its entirety, since it involves a number of now unnecessary steps, required at the time to ensure the purity of the inserted cDNA fragment, it was used to clone double-stranded cDNA fragments containing part of the nucleotide sequence of rat insulin mRNA (Ullrich *et al.*, 1977), rat growth hormone mRNA (Seeburg *et al.*, 1977a), human chorionic somatcmammotrophin mRNA (Shine *et al.*, 1977), mouse proopiocortin mRNA (Roberts *et al.*, 1979) and mRNA encoding the α-subunit of human chorionic gonadotrophin (Fiddes and Goodman, 1979). Consequently the strategy is worthy of brief description since it set experimental standards for others to follow.

The initial step was dependent on the ability of certain restriction enzymes (*Hae*III, *Hha*I) to cut single-stranded cDNA (see Blakesley and Wells, 1975) in addition to double-stranded cDNA, and so generate discrete fragments representative of the most abundant sequences within complex cDNA populations. Thus analysis of a ^{32}P-labelled human placental cDNA population after restriction with *Hae*III produced a discrete fragment of 550 ntd as judged by gel electrophoresis. Nucleotide sequence analysis of the 550 ntd fragment excised after gel electrophoresis, showed it to contain sequence complementary to an mRNA encoding chorionic somatomammotrophin (Seeburg *et al.*, 1977b). Once the identity of the 550 ntd fragment had been established, total double-stranded placental cDNA was synthesised, restricted with *Hae*III, the fragments separated on the basis of size by gel electrophoresis, and the 550 bp fragment eluted. To ensure purity, this was first dephosphorylated, then further restricted with *Hha*I into two fragments of 460 bp and 90 bp. These unique fragments — other contaminating 550 bp fragments were unlikely to have a *Hha*I site in a similar position — were again separated from each other and contaminating sequences by gel electrophoresis, eluted, then religated (only one orientation possible due to absence of 5′ phosphate groups at one end of each fragment).

The regenerated 550 bp fragment was rephosphorylated, and chemically synthesised *Eco*RI linkers attached by blunt end ligation, then the attached linkers cut with *Eco*RI to produce sticky ends. The 550 bp fragment with attached linkers was then annealed and ligated to *Eco*RI-cut pMB9 DNA, pretreated with alkaline phosphatase to prevent recircularisation of the plasmid in the absence of an inserted sequence. Transformation in this instance resulted in four recombinants, all identical. Positive identification of the inserted chorionic somatomammotrophin cDNA sequence was then established by DNA sequence analysis.

This strategy, while successful in its objectives at the time, requires an abundant mRNA species and results in the generation of a few recombinants containing the same sequence. Although any blunt ended double-stranded DNA fragment can be cloned via this route, other less abundant mRNA species are better cloned via an alternative route, since the use of *Eco*RI or *Hin*dIII, or any other suitable linkers (unless the appropriate methylase is available — for discussion see Williams, 1981) precludes the presence of internal restriction sites within the double-stranded cDNA if large fragments are to be cloned, unless nonsymmetrical linkers are used (Bahl *et al.*, 1978). Also rather obviously, the presence or absence of internal sites cannot be determined by restriction of the double-stranded cDNA population unless the sequence in question is abundant. Equally, positive correlation between prominent restriction enzymes fragments and a particular mRNA species cannot be made unless nucleotide sequence analysis can be performed.

B Synthesis of double-stranded cDNA and insertion into plasmid vectors using homopolymeric tails

A detailed and critical account of the construction of cDNA libraries has already been presented earlier in this series and should be referred to for many experimental considerations (Williams, 1981). Yet in a little over 18 months, techniques have improved, and options widen. The uses of linkers to construct cDNA-containing plasmids representative of abundant mRNA species has been superseded by strategies designed to create and store cDNA libraries representative of total cellular or tissue poly(A)-containing RNA populations, followed by selection of cDNA plasmids using sequence specific hybridisation probes. The construction of such libraries has generally involved the insertion of double-stranded cDNA into the chosen plasmid vector by means of complementary homopolymeric tails.

1 *Synthesis of double-stranded cDNA*

Two similar approaches, differing only in the method used for the synthesis of the second strand are depicted in Fig. 5. Both require the initial reverse transcription of poly(A)-containing RNA by AMV reverse transcriptase using an oligo dT_{12-18} primer. Assuming the integrity of the initial mRNA preparation, factors which influence full-length transcription include contaminating ribonucleases in the AMV reverse transcriptase preparation, or secondary structural constraints within the mRNA template itself which block efficient transcription. Ribonuclease degradation may be overcome either by further purifying the enzyme preparation (Retzel *et al.*, 1980), or more easily by the inclusion in the reaction mix of a simply purified RNase inhibitor originating from human placenta (Blackburn *et al.*, 1977). The latter has been used impressively to ensure the synthesis of full-length thyroglobulin cDNA of 8000 ntd (see Christophe *et al.*, 1982). Secondary structural constraints are more problematical, but to some extent may be overcome by prior denaturation of the poly(A)-containing RNA population using methyl mercury hydroxide (Payvar and Schimke, 1979), or brief heat treatment at 90°C in the presence of EDTA (Agarwal *et al.*, 1981). However, the possibility must be faced that certain sequences give rise to secondary structural constraints not easily overcome, thus preventing the synthesis of full length transcripts. We, for instance, have recently found that repetitive runs of $rG_{(3-4)}$ provide a formidable obstacle to reverse transcription (see Hall *et al.*, 1982).

Synthesis of the second cDNA strand after removal of the poly(A)-containing RNA by alkaline hydrolysis, and residual oligo$(dT)_{12-18}$ by gel filtration, may be obtained by two routes. One relies on the formation of a snap-back hairpin loop at the 3′ end of the single-stranded cDNA, thereby creating a primer for second strand synthesis. The other depends on the addition of a short homopolymer extension to the single-stranded cDNA, followed by second strand synthesis by AMV reverse transcriptase or a DNA polymerase, using the complementary oligodeoxynucleotide as the primer (Rougeon *et al.*, 1975). Once the second strand has been synthesised, incubation with S_1 nuclease removes the hair-pin loop if the first route is used, and degrades residual single stranded material (either route), leaving a double-strand cDNA population ready for the addition of homopolymeric tails and insertion into the chosen restriction site of the plasmid of choice. Seemingly simple, or should be — the technology is now well documented — provided that a few elementary precautions have been taken en route.

If the self-priming route for second strand synthesis is chosen, ensure that the final product is truly double-stranded — has the

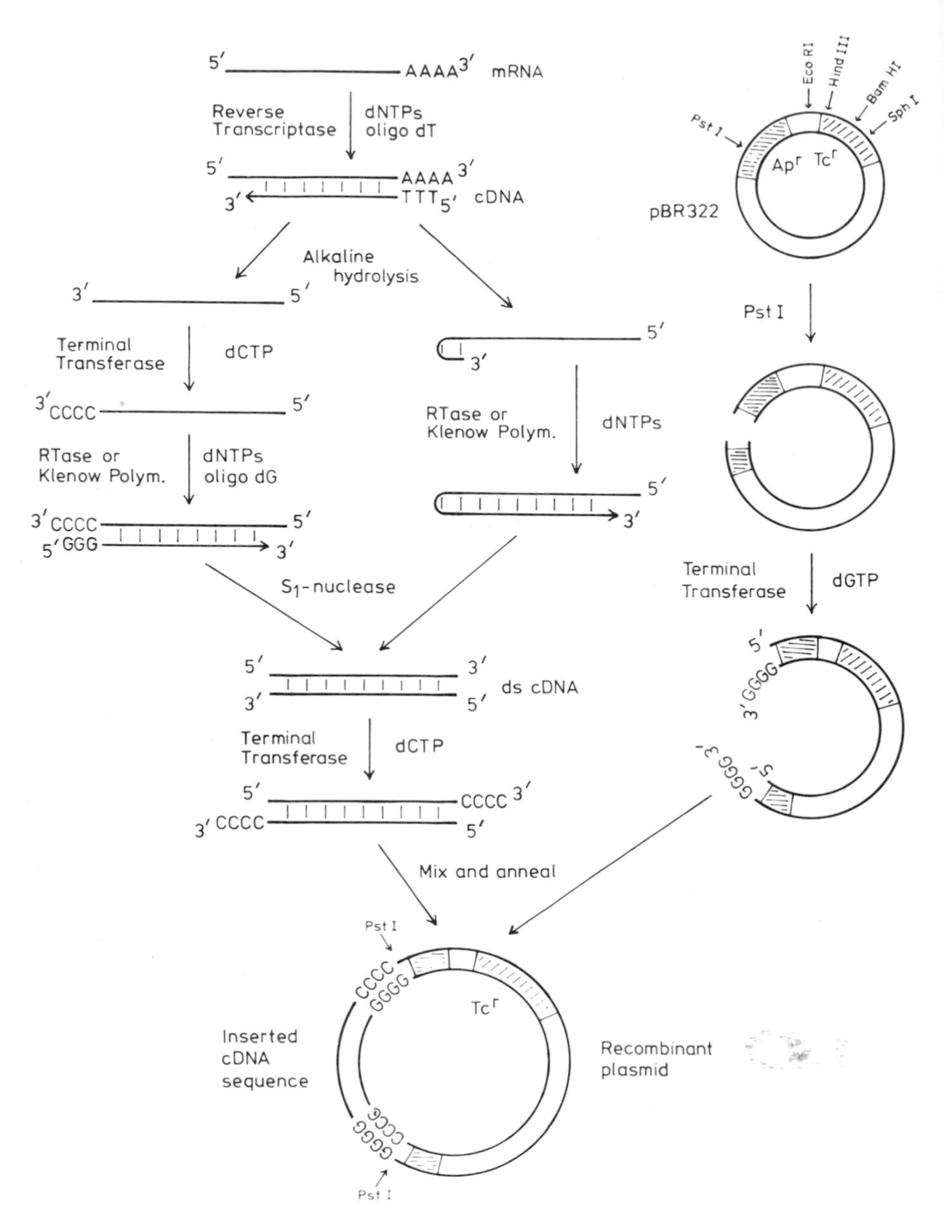

Figure 5 Synthesis of double-stranded cDNA and insertion into a plasmid vector. Alternative approaches, differing only in the method used for the synthesis of the second strand of the cDNA, are depicted. The self-priming route (which leads to recombinants lacking the extreme 5′-end of the mRNA), relies on the formation of a hair-pin loop at the 3′-end of the single-stranded cDNA, and is shown on the right. The tailing route (which may lead to full-length cDNA recombinants), involves the addition of a short homopolymeric extension to the 3′-end of the single-stranded cDNA, and is shown on the left. For further details see text.

hairpin been removed? If it has, heat denaturation of a small amount followed by rapid cooling will produce a single-stranded population, which will be totally degraded when incubated with low levels of S_1-nuclease. If the hair-pin remains, the strands will be resistant to S_1-nuclease digestion, indicating snap-back of the "pseudo" double-stranded molecule.

If the addition of homopolymeric tails to single-stranded cDNA using calf thymus terminal transferase is the chosen procedure, it is important to know with reasonable accuracy the number of available 3′ hydroxyl termini, since for a given amount of dXMP incorporated, the average length of the homopolymer added will be determined by the number of 3′ hydroxyl termini. This can be estimated by determination of the size profile of the single-stranded DNA population before the addition of the homopolymeric tails. If size determination by analytical agarose gel electrophoresis (McDonell *et al.*, 1977) or sucrose gradient centrifugation reveals a preponderance of low molecular weight cDNA "pieces" providing incalculable numbers of 3′ hydroxyl termini, then they should be removed, either by preparative sucrose gradient centrifugation or preparative agarose gel electrophoresis (see Craig *et al.*, 1981). The size distribution of the recovered material should then be determined, and a reasonable "guestimate" made of the number of 3′ hydroxyl termini present, based on the average size of the cDNA species. A similar size cut to remove low molecular weight material should also be performed on the final double-stranded cDNA preparation before the addition of homopolymeric tails, to remove residual oligonucleotide primer or cDNA fragments generated by S_1-nuclease digestion. It is at this stage that enrichment of specific sequences can be made on the basis of prior knowledge of the mRNA size (see section III.C), if the strategy is intended to clone a particular sequence as opposed to obtaining a representative cDNA library. Where nucleotide sequence of an mRNA is available, or can be predicted, then enrichment for specific sequences can be obtained during first strand synthesis by AMV reverse transcriptase using synthetic complementary oligodeoxynucleotide primers (see Villa-Komaroff *et al.*, 1978; Chan *et al.*, 1979; Noda *et al.*, 1982).

What are the merits of the two alternative strategies? For those new to these techniques, it is probably best to opt for the "hairpin" route, provided that a full-length clone is not essential for subsequent analysis, because this route will inevitably generate a preponderance of cDNA clones containing 3′ regions of the mRNA depending on the position of the hairpin. If cDNA clones containing the 5′ untranslated sequences or sequences encoding the

NH_2-terminal region of the protein are required, then the tailing route is best employed (see Land *et al.*, 1981, 1982).

2 *Homopolymeric tailing and site of insertion*

The addition of homopolymeric tails to the 3′ hydroxyl termini of single or double-stranded cDNA, or linearised plasmid, can be the cause of more aggravation than any other enzymic reaction during the construction of cDNA-containing recombinant plasmids. Provided that the starting double or single-stranded DNA is free of salt and/or traces of phenol which might inhibit the terminal transferase, and also of unwanted 3′ hydroxyl termini, then three factors are particularly important: the commercial source of terminal transferase, the length of the homopolymer tails; and the site of insertion into the vector, since this determines the choice of homopolymer.

The activity of commercial terminal transferase preparations vary, some in their ability to efficiently utilize each of the four different deoxyribonucleoside triphosphates, whilst others are active only over a short time course (10–20 min). If you have not used this enzyme before, don't take "pot-luck", find out from a cloning laboratory which source is currently the best; even some batches from the same commercial source are better than others.

Both cloning strategies outlined in Fig. 5 involve the addition of poly dC to the 3′ hydroxyl termini of the double-stranded cDNA population, and the addition of poly dG to the plasmid vector, in this instance pBR322 (Bolivar *et al.*, 1977), linearised by cleavage at the *Pst*I site. The advantages of restriction at this site are two-fold. First, since it lies within the gene which confers ampicillin resistance, inactivation of this gene by insertion of additional DNA sequence generally results in a change in phenotype which can be useful for the subsequent selection of recombinant plasmids. Secondly, dG tailing of the *Pst*I linearised plasmid generally (not always) results, after host repair, in restoration of *Pst*I sites at both ends of the inserted tailed cDNA sequence, thereby providing a means of excising the inserted sequence free of contaminating plasmid DNA (see Williams, 1981). Although insertion at the *Pst*I site is widely used, restoration of a restriction enzyme recognition site by homopolymer tailing is not limited to enzymes like *Pst*I and *Sph*I which cut leaving protruding 3′ hydroxyl termini. Other enzymes which cut leaving extended 5′ termini such as *Hin*dIII and *Bam*HI, and also lie within a gene confering drug resistance, may be used to linearise the plasmid DNA with subsequent restoration of the restriction site, provided that repair synthesis of plasmid DNA with T4 DNA polymerase has been carried out before tailing, and the appropriate deoxynucleoside triphosphates have been used to tail cDNA and vector DNA (see Deng and Wu, 1981). Once a decision has been made concerning the site of insertion into the plasmid vector, then the linearised vector (preferably purified by preparative gel electrophoresis to remove residual circular/supercoiled plasmid and ultimately the wild-type background in the final library), and the double-stranded cDNA preparation can be tailed with the appropriate homopolymer, annealed and transformed into the host bacteria.

A comprehensive analysis of the optimal tail lengths, annealing and transformation conditions for pBR322-derived recombinant plasmids into several *E. coli* host strains, has been described by Peacock *et al.* (1981). From these experiments it is clear that the optimum tail length for dA.dT is about 100 nucleotides, and that of dG.dC about 15—20 nucleotides. Such numbers can be difficult to attain in a controlled fashion, particularly when adding short lengths of homopolymers to small amounts of double-stranded cDNA (10—50 ng), since the enzyme may be inactivated when diluted, whilst too much enzyme can result in extensive polymerisation and thus overshoot in a matter of minutes. Recently it has been demonstrated that the reaction rate may be controlled under conditions which are substrate limiting (Deng and Wu, 1981). Using these procedures the addition of dG or dC tails of 15—20 ntd in length to small amounts of single or double-stranded cDNA can be achieved with considerable accuracy over a reasonable time course (30 min).

Most published experimental procedures involving the cloning of peptide hormone cDNA sequences using homopolymeric tails have utilised the "hairpin" route and dG.dC homopolymers in the *Pst*I site of pBR322 (see for instance Nakanishi *et al.*, 1978; Roskam and Rougeon, 1979; Hobart *et al.*, 1980; Hudson *et al.*, 1981a) though alternative sites and dA.dT homopolymers have also been used (see Harpold *et al.*, 1978; Chan *et al.*, 1979; Vamvakopoulos *et al.*, 1980; Gubler *et al.*, 1982). So far only in one reported instance has the "5′-tailing route" been taken (Land *et al.*, 1982). However, we anticipate that this may well become a favoured route, since the "interesting" end of most mRNA species in terms of protein coding potential lies towards the 5′ end.

3 Cloning efficiency

The strategies described above rely on a now well proven family of plasmid vectors using procedures which in technical terms are relatively simple. Until recently, through constraints imposed due to conjectural hazards, experiments involving the cloning of biologically active peptide sequences, even where expression was unlikely, have used safe "host-vector" systems as a means of reducing the required level of physical containment. This has generally meant the use of pBR322 transformed into the disabled *E. coli* strain χ1776 (Curtiss *et al.*, 1977), or alternatively, the use of non-mobilizable plasmids such as pAT153, a derivative of pBR322 (see Twigg and Sherratt, 1980), in *E. coli* recA^- strains, typically HB101 (Boyer and Roullard-Dussoix, 1969) — see Thompson (1982) for detailed discussion. Using these systems, published transformation frequencies above 20 colonies per ng input double-stranded cDNA have proved the exception rather than the rule. This has proved adequate for the cloning and selection of abundant mRNA sequences, but more efficient systems are required for the rapid generation of large cDNA libraries, particularly where the amount of poly(A)-containing RNA starting material is limited, and the sequence(s) of interest of low abundance. For instance a library of 100 000 colonies would be needed in order to obtain 10 clones containing cDNA sequences representative of an mRNA comprising 0.01% of the total poly(A)-containing RNA population.

We have found using the homopolymeric tailing procedures described by Deng and Wu (1981), and annealing and transformation procedures similar to those described by Peacock *et al.* (1981), that transformation efficiencies for a heterologous population of cDNA sequences inserted into the *Pst*I site using dG.dC tailing of 75—100 colonies per ng could be achieved using the pAT153/*E. coli* HB101 recA^- safe vector system, and that the number of colonies could be increased a further 2—3 fold (200—300 colonies per ng) if *E. coli* RR1, a recA^+ strain (Bolivar *et al.*, 1977), was used as the host strain. In the latter case 85—90% of the colonies obtained did not grow (or grew weakly) on ampicillin, an indication of the presence of inserted cDNA sequences. Even higher transformation efficiencies (2—3 fold) using RR1 have been reported by Peacock *et al.* (1981), using a defined DNA fragment to optimise conditions, whilst recently recombinants containing bovine enkephalin cDNA sequences (Gubler *et al.*, 1982) and rat relaxin cDNA sequences (Hudson *et al.*, 1981a) have been isolated from large cDNA libraries obtained by transformation of small amounts of chimaeric plasmid at high frequency into RR1. High transformation frequencies have also been attained using *E. coli* 294 (see Comb *et al.*, 1982b) and *E. coli* 5K (see Land *et al.*, 1982). The lower transformation frequencies obtained with *E. coli* K12 recA^- strains using homopolymer tailed chimaeric plasmids may reflect elevated exonuclease levels present in *E. coli* K12 recA^- strains (Williams *et al.*, 1981), since little difference in transformation frequency of supercoiled pBR322 DNA was reported when comparing K12 recA^+ and recA^- strains (Peacock *et al.*, 1981).

Using the technology described above, a realistic objective might be to generate in a matter of days a cDNA library of 20 000 colonies from $\leqslant 10\,\mu g$ of poly(A)-containing RNA as starting material. This would seem an adequate number of colonies provided that a source of tissue is readily available. However there remains room for improvement in instances where mRNA species of lower abundance are to be cloned and identified, and in situations where it might be desirable to generate cDNA libraries from restricted amounts of tissues or tumours, for example from a single animal. Recently a modified cDNA cloning procedure has been described in which the plasmid DNA vector tailed with oligo dT at one 3′ hydroxyl terminus, itself serves as a primer for first strand cDNA synthesis, then ultimately, after a series of modification steps, serves as a primer for the second strand synthesis. The procedure, though technically more demanding, provides greater cloning efficiency (10^5 plasmid-cDNA recombinants per μg mRNA — see Okayama and Berg, 1982), and has been successfully used to generate a library of 250 000 colonies from $3\,\mu g$ of porcine hypothalamic mRNA (Kakidani *et al.*, 1982). The development of high efficiency cloning systems such as these, shifts the technical emphasis away from the generation of "sufficient" clones, towards the techniques required for the efficient selection of required cDNA sequences from large libraries.

C Screening of cDNA libraries

Once cDNA libraries have been established, the identification and characterisation of individual plasmids containing specific cDNA sequences is carried out in two distinct phases. Firstly a broad screen of the complete library to identify colonies likely to contain recombinant plasmids of interest is carried out, followed by the detailed characterisation of the cDNA inserted into selected plasmids.

1 Primary screening

In the absence of expression, the strategies adopted reflect to a very great extent the procedures used to identify the mRNA species of interest within a given poly(A)-containing RNA population (see section III.C). Inevitably, preliminary screening of libraries utilize variations of the *in situ* hybridization procedure described by Grunstein and Hogness (1975) using radiolabelled RNA or DNA as hybridisation probes.

In situ hybridisation procedures are now well established. Where only several thousand colonies are to be screened, they should be picked in a grid pattern onto fresh agar containing the appropriate

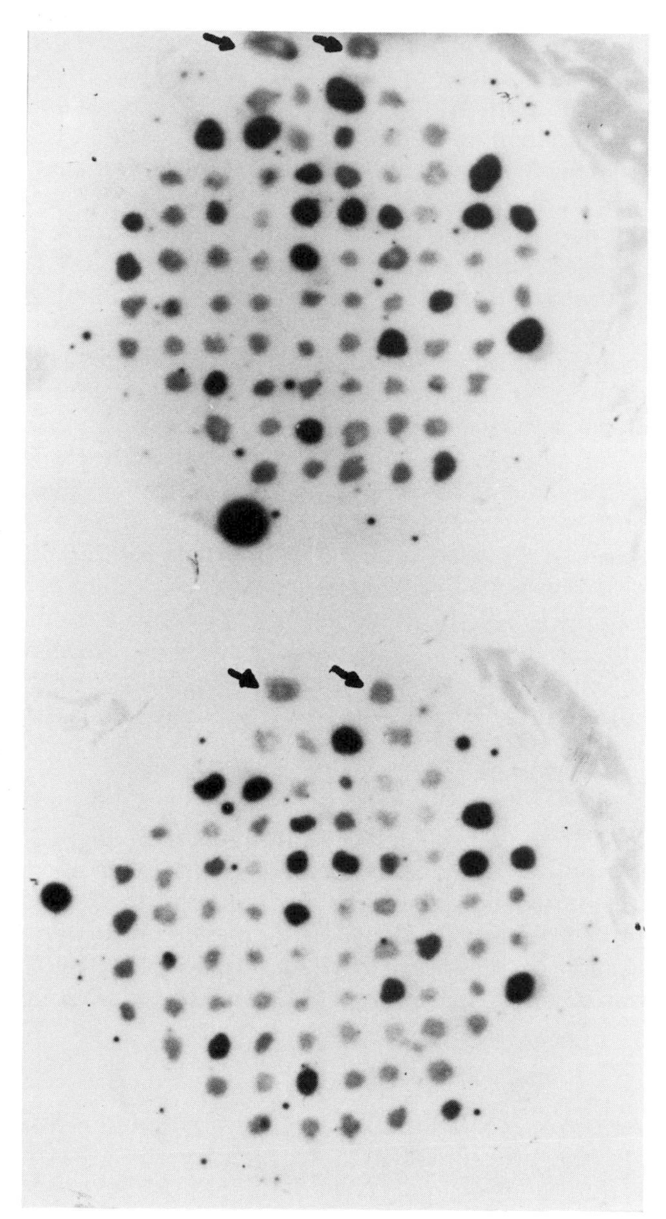

Figure 6A Primary screening of cDNA libraries. Analysis of the distribution of cloned abundant mRNA sequences within a cDNA library derived from human thyroid medullary carcinoma tissue. Recombinants from a human thyroid medullary carcinoma cDNA library (Allison *et al.*, 1981) were grown on nitrocellulose filters, and their DNA immobilised following lysis *in situ* with alkali. Colonies containing abundant cDNA sequences were then identified by hybridisation with ^{32}P-labelled base cleaved poly(A)-containing RNA which had been isolated from the thyroid medullary carcinoma. The two filters represent duplicates. Arrows indicate the colonies containing the parental plasmid pAT153.

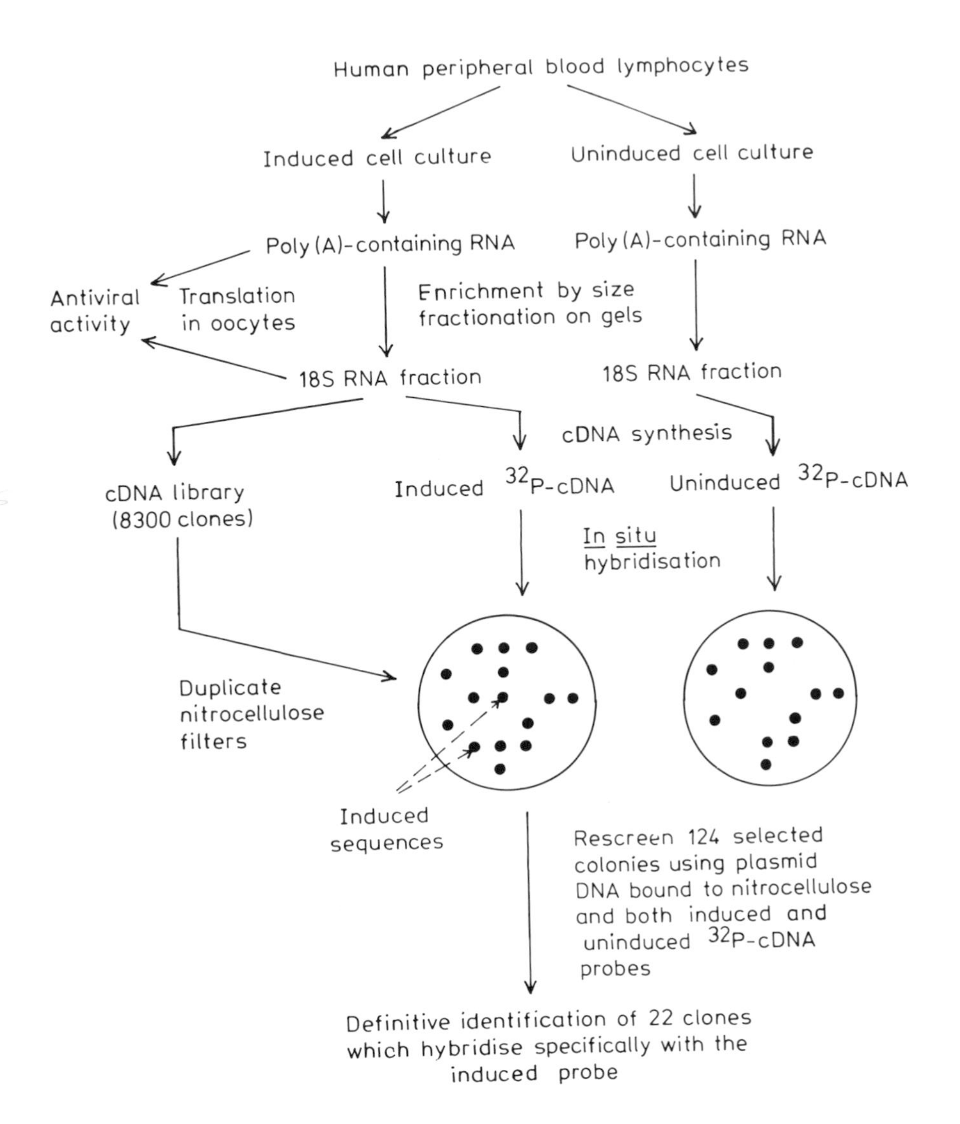

Figure 6B **Outline of the strategy used for the selection of recombinants containing human γ-interferon cDNA sequences (for details see section IV.C.1 and Gray *et al.*, 1982).**

antibiotic. This forms the master plate, from which each colony may then be replica plated onto nitrocellulose filters over-laid on agar dishes, grown, lysed with alkali, and the denatured DNA baked onto the filter (Thayer, 1979; also see Craig *et al.*, 1981; Williams, 1981). Where large numbers of colonies are involved, the adapted method for high colony density of Hanahan and Meselson (1980) should be followed (see also Dahl *et al.*, 1981). Once the colonies have been fixed, the filters may then be screened using a variety of radiolabelled hybridisation probes, dependent on the individual circumstance.

An example of the use of a total ^{32}P-labelled poly(A)-containing RNA preparation for the preliminary selection of cDNA plasmids containing human calcitonin precursor cDNA sequences is shown in Fig. 6A. In this instance the mRNA was known to be abundant as judged by cell-free translation (see Fig. 2). Moreover, to increase the specific activity of the hybridisation probe, the number of 5′ phosphate termini available for exchange labelling with [γ-^{32}P] ATP and T4 polynucleotide kinase was increased by limited hydrolysis of the RNA with alkali, resulting in a final specific activity of 2×10^7 cpm/μg (see Craig *et al.*, 1981). A similar approach using an enriched mRNA preparation has been used to identify, in a preliminary manner, plasmids containing cDNA sequences encoding the bovine proopiocortin mRNA (Nakanishi *et al.*, 1978) and the bovine arginine vasopressin-neurophysin II precursor mRNA (Land *et al.*, 1982). Others have used ^{125}I-labelled mRNA (Vamvakopoulos *et al.*, 1980) or even ^{32}P-labelled cRNA synthesised from the initial rat growth hormone-enriched cDNA preparation (Harpold *et al.*, 1978). ^{32}P-labelled cDNA prepared from total mRNA may be also used for preliminary screening if the sequence of interest is abundant (Roskam and Rougeon, 1979; Cooke *et al.*, 1981; Hobart *et al.*, 1980; Keshet *et al.*, 1981). Alternatively, the relative abundance of a cDNA hybridization probe can be increased, by synthesis from an enriched mRNA preparation (Nilson *et al.*, 1980).

If the mRNA sequence of interest is of relatively low abundance, or poly(A)-containing RNA is scarce, the following selection procedures should be considered. The first takes advantage of homologies between mRNA encoding the same peptide hormone in different species. Using this approach Bell *et al.* (1979) identified a human preproinsulin cDNA clone using a nick-translated cloned rat preproinsulin I cDNA probe (see Ullrich *et al.*, 1977), and Hendy *et al.* (1981) identified a human preproparathyroid hormone cDNA clone using a nick-translated cloned bovine preproparathyroid hormone cDNA probe (see Kronenberg *et al.*, 1979). The second selection procedure utilises chemically synthesised oligodeoxynucleotides either of known, or predicted nucletide sequence, essentially as described in the previous section for the identification of specific poly(A)-containing RNA species (Section III.C). The primers may either be ^{32}P-labelled directly with [γ-^{32}P]ATP and T4 polynucleotide kinase, or alternatively may be used to prime the synthesis of ^{32}P-labelled cDNA from a total or enriched poly(A)-containing RNA population. Radiolabelled oligodeoxynucleotides have been used directly to select bovine (Gubler *et al.*, 1982; Noda *et al.*, 1982) and human (Comb *et al.*, 1982b) adrenal preproenkephalin cDNA clones, and porcine hypothalamic preproenkephalin cDNA clones (Kakidani *et al.*, 1982). Gel purified ^{32}P-labelled cDNA, synthesised by sequence specific primer extension, has been used to identify rat and porcine preprorelaxin (Hudson *et al.*, 1981a; Haley *et al.*, 1982) and rat preproinsulin (Chan *et al.*, 1979) cDNA clones.

A third procedure involves the differential screening of cDNA libraries for relatively low abundance mRNA species encoding biologically active peptides of unknown amino-acid sequence. One of the best published examples to date describes the identification of human γ-interferon cDNA sequences (see Gray *et al.*, 1982). In quantitative terms this involved the cloning and selection of an mRNA sequence representing $\sim 0.04\%$ of the poly(A)-containing RNA from human peripheral blood lymphocytes induced in culture to produce γ-interferon. The strategy required that mRNA directed protein synthesis gave rise to a biologically active peptide of measurable activity, in this instance by injection of total poly(A)-containing RNA into *Xenopus* oocytes and subsequent identification of antiviral activity of the secreted products in the oocyte incubation medium. Given that a bioassay could be performed, the γ-interferon mRNA was first enriched (20-fold) by size fractionation, and the resulting poly(A)-containing RNA (18S) used as the template for the synthesis of ^{32}P-labelled cDNA. However, since the "enriched" poly(A)-containing RNA still consists predominantly of normal lymphocyte poly(A)-containing RNA species, a size fractionation of normal (uninduced) lymphocyte poly(A)-containing RNA was also performed and a corresponding ^{32}P-labelled cDNA population prepared. "Induced" and "uninduced" cDNA probes were each then used to probe by *in situ* hybridisation an induced "enriched" lymphocyte cDNA library (8300 transformants) which had been grown up and lysed on two separate sets of filters (see Fig. 6B). Comparison of the resultant autoradiographs revealed 124 colonies which hybridised with the induced probe, but only weakly or not at all with the uninduced probe. Rescreening of each of these "positive" colonies, in this instance using purified plasmid DNA bound to nitrocellulose, showed that of the selected colonies, 22 hybridised specifically to the induced probe. Further characterisation of the inserted cDNA showed that all contained related sequences. Characterisation of the largest inserted cDNA sequence showed it to contain a complete γ-interferon cDNA sequence as judged by nucleotide sequence determination, and the ability of the cloned cDNA to direct the synthesis of a peptide with the expected antiviral activity when inserted in an expression vector. Differential hybridisation should prove generally useful for the primary selection from cDNA libraries of low abundance mRNA species encoding biologically active peptides such as growth factors where parallel screening can eliminate normal cellular mRNA sequences of similar size and abundance present in both induced and uninduced state, or common to all cells and tissues.

In situations where expression of cDNA sequences within a cDNA library has been sought, then primary screening can be performed using radiolabelled antibodies. Thus immunological screening procedures have been employed to identify colonies in which inserted insulin cDNA sequences were expressed (Broome and Gilbert, 1978; see also Dahl *et al.*, 1981; Williams, 1981).

2 *Secondary screening*

Secondary screening of candidate plasmids from cDNA libraries requires the identification of the inserted cDNA sequence such that nucleotide sequence determination may be performed with confidence. As a preliminary to this, selected colonies should be cultured on a small scale, the plasmid DNA isolated using rapid procedures (see Birnboim and Doly, 1979; Holmes and Quigley, 1981), and the size of the inserted cDNA sequence determined by horizontal agarose gel electrophoresis (see McDonell *et al.*, 1977). Ideally, if the design of the recombinants results in reconstruction of restriction sites at both ends of the inserted sequence (see section IV.B.2), then the size of the insert may be determined after restriction by comparative electrophoresis with restriction fragments of known size (see Sutcliffe, 1978a, b; also Fig. 7A). Subsequent screening procedures reflect to a very great extent the various factors considered in the earlier section designed to identify the presence of a given mRNA species within a particular poly(A)-containing RNA population.

Where an amino-acid sequence is available, unambiguous identification of an inserted cDNA sequence requires nucleotide sequence analysis. Where sequence-specific hybridisation probes have been utilised for the preliminary selection, plasmids containing the largest inserted cDNA sequences should be subjected to DNA sequence analysis immediately. Similarly, where recombinants have been selected using immunological techniques to detect an expressed product (see Broome and Gilbert, 1978), then nucleotide sequence determination should proceed without further screening. If, on the other hand, large numbers of recombinants are still being handled, for example after using a "mixed" probe representing several abundant mRNA species for the initial *in situ* screening, then a "weeding-out" procedure is preferable before DNA sequence analysis is undertaken. The most commonly used secondary screening method, positive mRNA hybrid selection, requires immobilisation of "nicked" denatured plasmid DNA on a nitrocellulose or diazotized paper support (see Williams, 1981, for detailed considerations). Incubation of these with the appropriate poly(A)-containing RNA population using suitable ionic conditions results in the formation of a plasmid cDNA-mRNA hybrid. After suitably stringent washing procedures have been performed to remove unhybridised poly(A)-containing RNA, the hybridized sequence may be eluted, and the protein it encodes identified using a combination of cell-free protein synthesis and SDS/polyacrylamide gel electrophoresis (see Fig. 7B). This procedure provides a means of differentiating between recombinant plasmids encoding different mRNA species of similar abundance within the same population, and together with the preliminary size analysis, permits selection of plasmids containing the largest inserted cDNA sequences representative of each mRNA species for subsequent DNA sequence analysis. However, it is only readily suited to systems where the mRNA species are relatively abundant (at least 0.2–0.5%), except where a low abundance mRNA directs the synthesis and secretion in oocytes of a biologically active peptide as opposed to an inactive precursor protein. In such a situation, where the initial *in situ* colony hybridisation may have proved indeterminate, or several hundred potentially interesting recombinants have been identified, colonies may be pooled in batches (20–50), the plasmid DNA isolated, immobilised on filters, then incubated with total or "enriched" poly(A)-containing RNA sequences. Hybridised mRNA is then eluted, concentrated by ethanol precipitation, microinjected into oocytes, and the oocyte incubation medium assayed for the required activity. Once activity within a given batch has been identified, all colonies within the same batch may then be screened individually, and the plasmid(s) containing the cDNA sequence(s) of interest identified. When a single recombinant has been identified the inserted sequence can then be excised, ^{32}P-labelled and used to reprobe the library (see Weissmann, 1981, for an application to the identification of interferon cDNA recombinant plasmids). As described in section III.C, granulocyte-macrophage colony stimulating factor activity has been identified in oocyte secretions as a result of microinjected mRNA (see Lusis *et al.*, 1982). No doubt

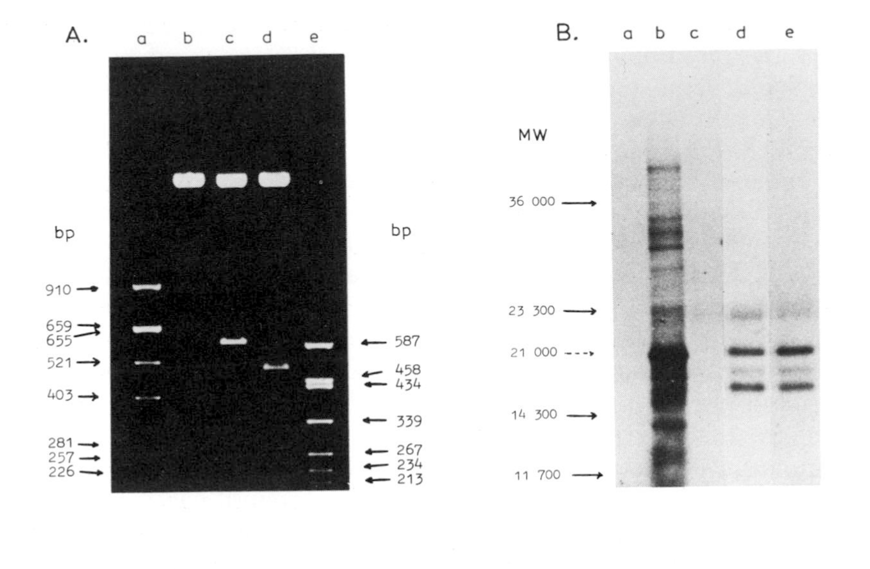

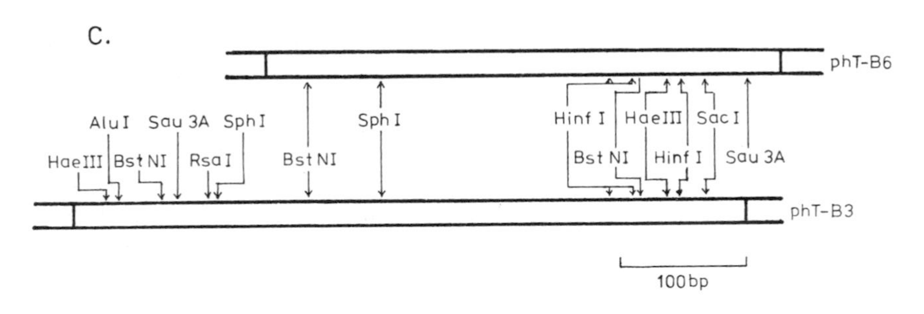

Figure 7 Characterisation of plasmids containing human calcitonin precursor polyprotein cDNA sequences. Two of the recombinant plasmids derived from a human thyroid medullary carcinoma cDNA library (Allison *et al.*, 1981) and shown to contain abundant mRNA sequences (see Fig. 6) were subjected to further analysis. These plasmids, designated phT-B3 and phT-B6, were constructed by inserting cDNA sequences into the *Pst*I site of pAT153, using dG.dC homopolymeric tails (Allison *et al.*, 1981). A. Size analysis of inserted cDNA sequences. Plasmid DNA samples (0.25 μg) were digested with restriction endonucleases, then electrophoresed on a 1.6% agarose gel (McDonell *et al.*, 1977) in the presence of ethidium bromide. Tracks (a) and (e), DNA fragments of known size (Sutcliffe, 1978a, b) generated by *Alu*I digestion of pBR322 and *Hae*III digestion of pAT153 respectively. Tracks (b), (c) and (d), *Pst*I digested pAT153, phT-B3 and phT-B6, respectively. B. Positive hybridisation translation. Sequence-specific mRNA isolated by hybridisation to denatured recombinant DNA, immobilized on DBM-paper (Craig *et al.*, 1981), was translated in the wheat-germ cell-free system as described in Fig. 2 (see Allison *et al.*, 1981). Track (a), no added RNA; (b), total human thyroid medullary carcinoma mRNA;

poly(A)-containing RNA species encoding this and other growth factors will soon be cloned and the appropriate recombinant cDNA plasmids selected from large cDNA libraries using strategies similar to those developed for the cloning and selection of recombinant plasmids containing interferon cDNA sequences.

In the absence of sequence specific hybridisation probes or suitable amounts of homologous poly(A)-containing RNA, necessary for positive mRNA hybrid selection procedures, two alternatives may be considered. First, it is possible that poly(A)-containing RNA obtained from the same tissue in other related species may show sufficient nucleotide sequence homology to permit primary and secondary screening of the cDNA clones of interest, particularly if the stringency of the hybridisation conditions is reduced (see Hall *et al.*, 1981). Secondly, prior knowledge of amino-acid sequence can in certain instances be used to predict the presence of restriction enzyme sites. This then permits the identification of cDNA sequences by the presence of a unique restriction fragment of predicted size within the inserted sequence. This approach has been used to identify cDNA clones containing the coding sequence of the β-subunit of human chorionic gonadotrophin (Fiddes and Goodman, 1979), and the bovine arginine-vasopressin-neurophysin II precursor (Land *et al.*, 1982). In each instance the strategy was based on the predicted presence of a *Sau* 96I or *Asu* I site (GGNCC) wherever the dipeptide Gly-Pro (GGNCCN) occurred. An alternative example, the dipeptide Ala-Ala (GCNGCN) would indicate the presence of an *Fnu* 4HI site (GCNGC). Thus provided that the number of amino acids separating the chosen pair of dipeptides is known, then the size of the corresponding cloned cDNA fragment may be predicted and the plasmid identified. However, the results should be interpreted with caution since the chosen restriction enzymes may cleave at other positions possibly between the predicted sites and in a different "reading" frame, thus complicating interpretation of the final result.

D Definitive identification of selected sequences by nucleotide sequence determination

Conclusive identification of an inserted cDNA sequence, where the amino-acid sequence is known, rests with determination of the

---

(c), (d) and (e), specific mRNA isolated by hybridisation to plasmids pAT153, phT-B3 and phT-B6 respectively. C. Comparative restriction endonuclease maps of the inserted cDNA within recombinant plasmids phT-B3 and phT-B6 confirms that they contain common sequence (reprinted with permission of the Biochemical Journal).

nucleotide sequence, and identification of the correct reading frame. DNA sequence analysis of selected inserted cDNA sequences is generally undertaken using one of two methods – the chemical cleavage procedure of Maxam and Gilbert (1977) or the chain termination procedure of Sanger *et al.* (1977). The chemical cleavage procedure has the advantage that, provided a rudimentary restriction map of the inserted cDNA sequence has been generated, nucleotide sequence determination may proceed immediately (see Fig. 8). In contrast the chain termination procedure, although requiring less manipulative procedures to generate the final DNA sequencing ladder, is based on an enzymatic reaction, requiring the extension of a defined oligodeoxynucleotide primer by a DNA polymerase along a single-stranded DNA template (see Smith, 1980, for detailed consideration of sequence analysis by primed synthesis). Until recently, the isolation of the primer and template provided a time consuming stumbling-block to the general application of this procedure. However, the construction of a series of cloning vehicles based on M13, a single-stranded DNA coliphage, which has a double-stranded replicative intermediate, has largely overcome these problems, since the double-stranded form can be treated as a plasmid vector. This provides a means by which double-stranded cDNA fragments may be excised by restriction endonuclease digestion from selected plasmids, inserted into restricted replicative intermediate M13 DNA by ligation, either directly or using linkers, and transformed into the appropriate *E. coli* host strain. Three features of the M13 cloning vehicles are particularly attractive (see Fig. 8). First, they incorporate a region of the *lac* promoter-operator resulting in expression of part of the β-galactosidase gene, sufficient to complement certain *E. coli* K12 strains deficient in β-galactosidase activity. This establishes a selection procedure since the insertion of additional DNA within this region results in the loss of complementation, giving rise to colourless as opposed to blue plaques on selective agar containing X-gal (see Dahl *et al.*, 1981). Secondly, the insertion of a multipurpose cloning site, a region of DNA within the *lac* promoter-operator containing multiple restriction enzyme sites (*Eco*RI, *Sal*I, *Acc*I, *Bam*HI, *Hinc*II and *Pst*I), widens the range of restriction fragments which may be recloned directly, or indirectly using synthetic linkers (see Williams, 1981), effectively increasing the versatility of the vector. Thirdly, by cloning within a defined region of the phage, only a single "universal" primer complementary to adjacent *lac* DNA is required for the initial nucleotide sequence determinations, whatever the sequence of the inserted cDNA fragment (see Anderson *et al.*, 1980; Heidecker *et al.*, 1980).

At present, the chemical cleavage method probably remains the most commonly used procedure, since it avoids the necessity of recloning into M13. Excellent detailed protocols for the chemical cleavage procedure covering sequencing strategies, the isolation and labelling of specific restriction fragments, nucleotide sequence determination and hints for trouble-shooting have been described by Maxam and Gilbert (1980). We would add to this a few comments which may prove helpful to those embarking on nucleotide sequence determination for the first time. Much emphasis has been placed on the use of T4 polynucleotide kinase and [α-^{32}P]ATP to label the 5′ termini of linearised plasmid or purified restriction fragments. These protocols in our hands proved highly efficient when labelling 5′ protruding ends (e.g. *Eco*RI, *Hind*III cut fragments), but were less efficient when fragments containing flush or recessed 5′ termini were used. Moreover, the efficiency of direct nucleotide sequence analysis of whole plasmid DNA, cut at single or multiple restriction sites within the inserted cDNA sequence (as opposed to specific restriction fragments purified by acrylamide gel electrophoresis) and radiolabelled by this means, is adversely effected by residual RNA preferentially radiolabelled in the reaction. These problems can be overcome by labelling the 3′ hydroxyl termini by one of two methods. Either cordycepin 5′-[α-^{32}P] triphosphate, a chain terminator, may be used as substrate for calf thymus terminal transferase (Tu and Cohen, 1980), or alternatively T4 DNA polymerase with its associated 3′→5′ exonuclease activity may be used as a means of labelling 3′ hydroxyl termini, whether these have been generated as recessed, flush or protruding ends (see Challberg and Englund, 1980). Both systems have the advantage over T4 polynucleotide kinase that plasmid DNA preparations contaminated with RNA may be used since the enzymes are inactive on RNA substrates. However, T4 DNA polymerase can have an additional advantage, in that not only can any 3′ hydroxyl termini be radiolabelled, but careful choice of [α-^{32}P]deoxynucleoside triphosphate can result in the selective labelling of one end only of a restriction fragment, avoiding the necessity of secondary restriction, or strand separation procedures, before the chemical cleavage reactions can take place (see Fig. 8). A hypothetical example of this might be a fragment generated by a *Hpa*II/*Rsa*I double digest:

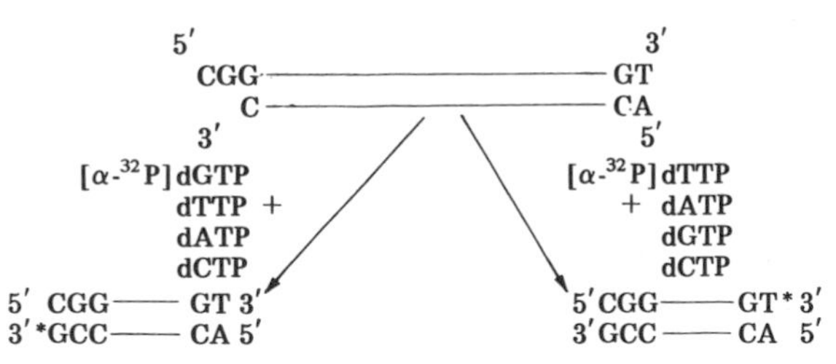

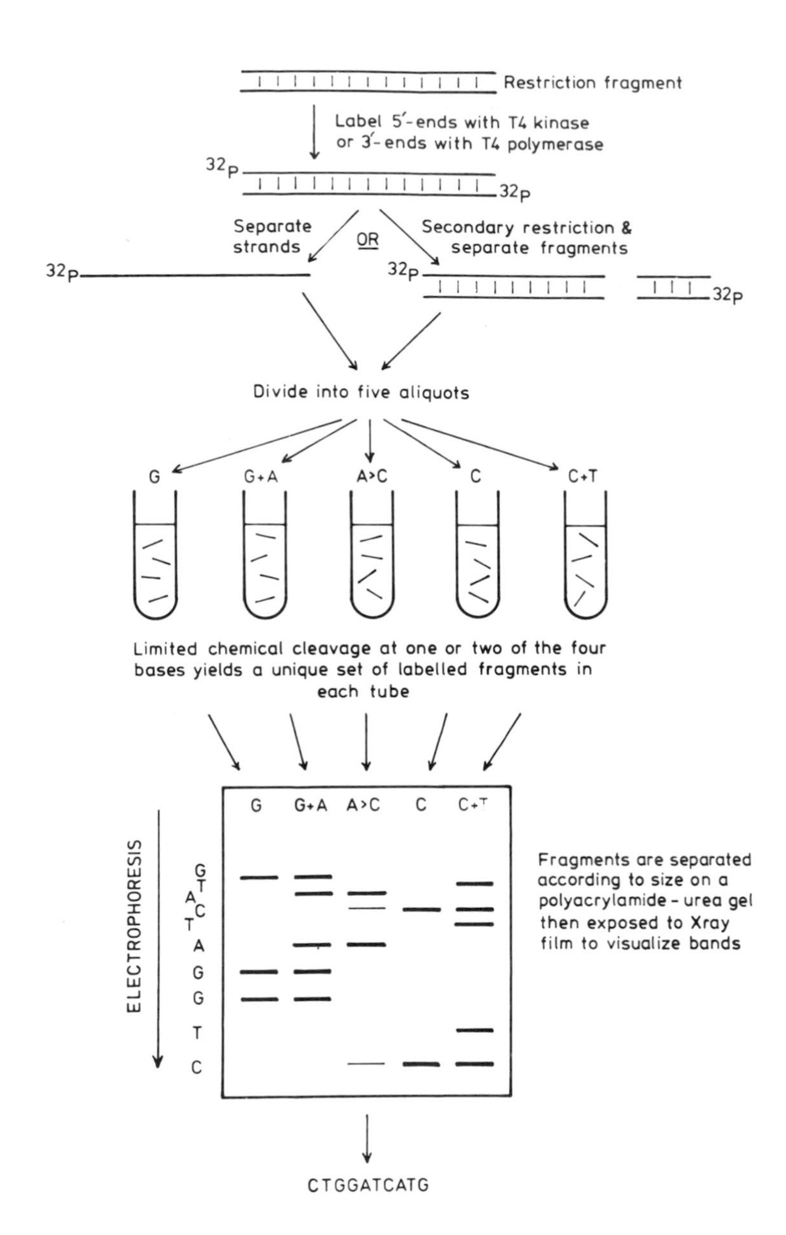

Figure 8 DNA sequence determination. Outline of the steps involved in the chemical cleavage procedure of Maxam and Gilbert (1977) using terminally labelled single or double-stranded DNA fragments (above), and in the chain termination procedure of Sanger *et al.* (1977) as applied to cDNA subcloned into M13 (opposite).

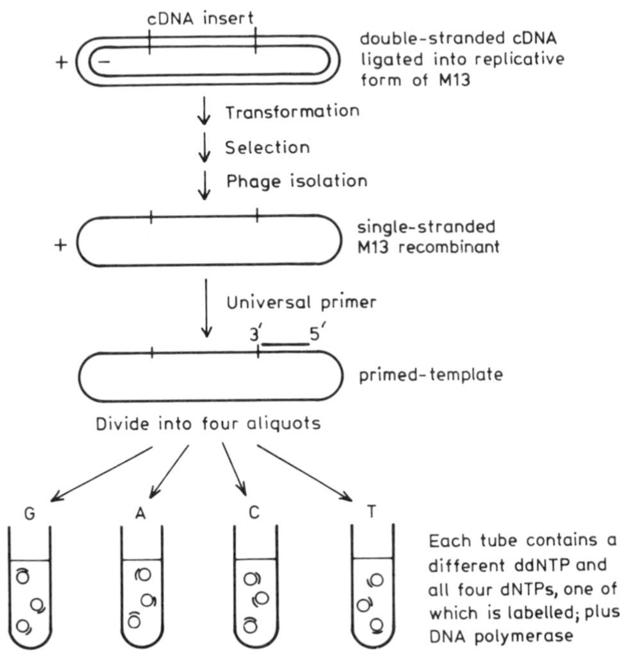

DNA synthesis is terminated randomly by incorporation of a ddNTP to yield a unique set of labelled fragments in each tube.

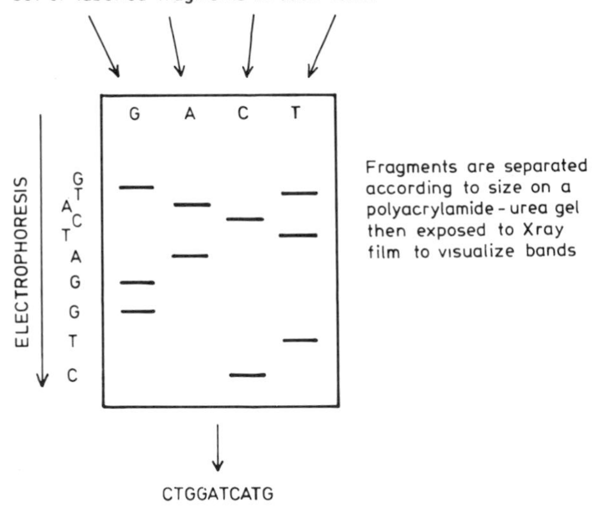

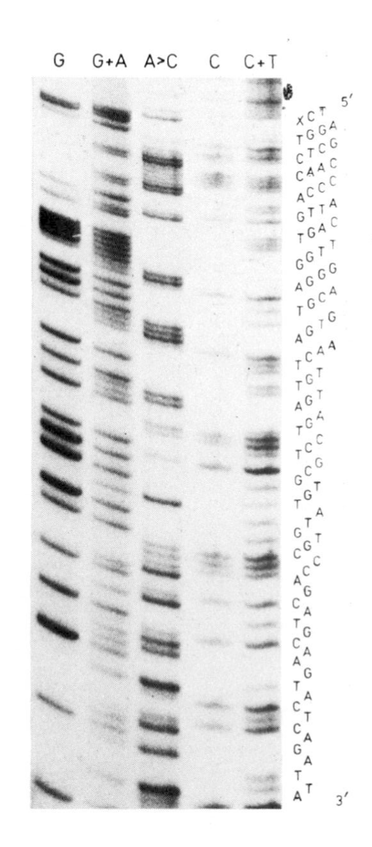


```
-36 -35 -34 -33 -32 -31 -30 -29 -28 -27 -26 -25 -24 -23 -22 -21 -20 -19 -18 -17 -16 -15 -14 -13 -12 -11 -10 -9
Val Leu Leu Ala Ala Leu Val Gln Asp Tyr Val Gln Met Lys Ala Ser Glu Leu Glu Gln Glu Gln Glu Arg Glu Gly Ser Ser

5' GTC CTG CTG GCT GCA CTG GTG CAG GAC TAT GTG CAG ATG AAG GCC AGT GAG CTG GAG CAG GAG CAA GAG AGA GAG GGC TCC AGC
3' CAG GAC GAC CGA CGT GAC CAC GTC CTG ATA CAC GTC TAC TTC CGG TCA CTC GAC CTC GTC CTC GTT CTC TCT CTC CCG AGG TCG

-8  -7  -6  -5  -4  -3  [-2  -1 ] 1   2   3   4   5   6   7   8   9   10  11  12  13  14  15  16  17  18  19  20  21  22  23
Leu Asp Ser Pro Arg Ser [Lys Arg] Cys Gly Asn Leu Ser Thr Cys Met Leu Gly Thr Tyr Thr Gln Asp Phe Asn Lys Phe His Thr Phe Pro

CTG GAC AGC CCC AGA TCT AAG CGG TGC GGT AAT CTG AGT ACT TGC ATG CTG GGC ACA TAC ACG CAG GAC TTC AAC AAG TTT CAC ACG TTC CCC
GAC CTG TCG GGG TCT AGA TTC GCC ACG CCA TTA GAG TCA TGA ACG TAC GAC CCG TGT ATG TGC GTC CTG AAG TTG TTC AAA GTG TGC AAG GGG
                                        <--------------------------------------------------------------------------------

24  25  26  27  28  29  30  31  32  [+1 ][+2  +3  +4 ] +5  +6  +7  +8  +9  +10 +11 +12 +13 +14 +15 +16 +17 +18 +19 +20 +21 +22
Gln Thr Ala Ile Gly Val Gly Ala Pro [Gly][Lys Lys Arg] Asp Met Ser Ser Asp Leu Glu Arg Asp His Arg Pro His Val Ser Met Pro Gln

CAA ACT GCA ATT GGG GTT GGA GCA CCT GGA AAG AAA AGG GAT ATG TCC AGC GAC TTG GAG AGA GAC CAT CGC CCT CAT GTT AGC ATG CCC CAG
GTT TGA CGT TAA CCC CAA CCT CGT GGA CCT TTC TTT TCC CTA TAC AGG TCG CTG AAC CTC TCT CTG GTA GCG GGA GTA CAA TCG TAC GGG GTC
---------------------------------------->

+23 +24 +25
Asn Ala Asn term

AAT GCC AAC TAA ACTCCTCCCTTTCCTTCCTAATTTCCCTTCTTGCATCCTTCCTATAACTTGATGCATGTGGTTTGGTTCCTCTCTGGTGGCTCTTTGGGCTGGTATTGGTCGCTTTC
TTA CGG TTG ATT TGAGGAGGGAAAGGAAGGATTAAAGGGAAGAACGTAGGAAGGATATTGAACTACGTACACCAAACCAAGGAGAGACCACCGAGAAACCCGACCATAACCACCGAAAG

CTTGTGGCAGAGGATGTCTCAAACTTCAAGATGGGAGGAAAGAGAGCAGGACTCACAGGTTGGAAGAGAATCACCTGGGAAAATACCAGAAAATGAGGGCCGCTTTGAGTCCCCCAGAGATGT
GAACACCGTCTCCTACAGAGTTTGAAGTTCTACCCTCCTTTCTCTCGTCCTGAGTGTCCAACCTTCTCTTAGTGGACCCTTTTATGGTCTTTTACTCCCGGCGAAACTCAGGGGGTCTCTACA

CATCAGAGCTCCTCTGTCCTGCTTCTGAATGTGCTGATCATTTGAGG[AATAAA]ATTATTTTTCCCC(A)n 3'
GTAGTCTCGAGGAGACAGGACGAAGACTTACACGACTAGTAAACTCC[TTATTT]TAATAAAAAGGGG(T)n 5'
```

Inclusion of [α-^{32}P]TTP and the remaining three "cold" deoxyribonucleoside triphosphates would give rise to preferential exchange radiolabelling of the upper strand, due to a combination of exonuclease $3' \rightarrow 5'$ activity and repair DNA synthesis by the polymerase, whilst the lower strand would be gap-filled and remain unlabelled. Conversely the inclusion of [α-^{32}P]dGTP and the remaining three appropriate "cold" deoxynucleoside triphosphates would result in radiolabelling of the *Hpa*II site only, the lower strand, in this example. In certain instances where limited sequence data is already known at each end of restriction fragments generated by enzymes such as *Hinf*I (G ↓ ANTC) then the same strategy in certain situations, may be used giving rise to the preferential radiolabelling of one strand only of one of a mixture of restriction fragments, again avoiding additional restriction and gel separation/elution procedures.

Once fragments labelled at a single terminus of one strand only have been generated, it is advisable to perform five separate partial cleavage reactions (see Figs 8 and 9), before the sequence ladders are analysed by electrophoresis on thin polyacrylamide-urea gels (Sanger and Coulson, 1978). The inclusion of the A > C cleavage reaction we find a useful safe-guard. The reaction itself is simple, and can be particularly useful should residual salt create problems in one or both pyrimidine tracks (C or C + T). Overall, DNA sequence determination using this method can be a rapid and simple procedure, provided that the DNA used throughout is free of salt and other chemical contaminants (e.g. phenol). Both can cause havoc at any step, from the preliminary restriction mapping through to the final cleavage reactions of radiolabelled DNA.

What if full length cDNA clones are not obtained? Then provided that the cloned DNA lacks sequence corresponding to the 5′ region

Figure 9 Partial nucleotide sequence of human calcitonin precursor polyprotein mRNA. The sequence depicts the total nucleotide sequence of the inserted cDNAs within recombinant plasmids phT-B3 and phT-B6 (see Fig. 7 and Craig *et al.*, 1982), determined by the method of Maxam and Gilbert (1977). The deduced amino-acid sequence confirms the presence of human calcitonin (residues 1 to 32), but in addition reveals the presence of amino terminal (− 1 to − 36) and carboxyl terminal (+ 1 to + 25) flanking cryptic peptides. Boxed sections on each side of the calcitonin sequence (− 2, − 1 and + 2 to + 4) indicate basic amino acids which are probably sites for proteolytic processing. The glycine residue (+ 1) immediately adjacent to the carboxyl-terminal proline of calcitonin fulfills the requirement of an amidated proline in the fully processed hormone (see section II.A). The underlined nucleotide sequence (from amino-acid residues 2 to + 1), which contains all but the first two amino acids of mature calcitonin, can be read from the sequencing ladder shown on the right (in this case, the complementary sequence to the mRNA). The "X" residue, three nucleotides from the top of the sequence, indicates a gap in the ladder characteristic of a methylated cytosine residue (see Ohmori *et al.*, 1978).

of the mRNA, additional sequence may be obtained by primer directed synthesis as outlined previously (section III.C) using a poly(A)-containing RNA template, and either a synthetic oligodeoxynucleotide primer, or a primer derived from a restriction fragment (see also Smith , 1980; Ghosh *et al.*, 1980).

Whichever method of DNA sequence determination is chosen, it is imperative that both strands of the inserted double-stranded cDNA be sequenced. Only if this has been performed, and it has been established that one strand is complementary in its entirety to the other, can the primary amino-acid sequence encoded within the mRNA be deduced with confidence and other structural features identified. Generally, with the exception of Glu/Gln and Asp/Asn (pairs of amino acids which are difficult to distinguish from each other using protein sequencing techniques), DNA sequence determination of cloned cDNA sequences has confirmed previously reported primary amino-acid sequence of the large multi-chain glycoprotein and large single-chain peptide hormones and has, in some instances, clarified amino-acid sequence which has been in dispute (see Kronenberg *et al.*, 1979). Sequence analysis of full length cDNA sequences has also defined the amino acid sequence of the NH_2-terminal signal peptides, found to be encoded by all peptide hormone mRNA species so far examined, has defined the relative position of the peptide(s) within precursor polyproteins (see Nakanishi *et al.*, 1979), and has identified in some instances previously unknown flanking cryptic peptides as judged by the position of paired basic amino-acid residues. For instance, nucleotide sequence determination of part of the human calcitonin cDNA sequence (Fig. 9, see also Craig *et al.*, 1982) identifies the region encoding human calcitonin (32 amino acids plus an additional carboxyl-terminal glycine) and shows that this peptide is separated from previously unknown flanking cryptic peptide sequences by basic amino-acid residues. The presence of the glycine residue immediately adjacent to the carboxyl-terminal proline in the calcitonin precursor reflects the requirement of an amidated carboxyl-terminal proline in the fully processed calcitonin, a situation common to other small peptides containing carboxyl-terminal amidated amino-acid residues, and known to be synthesized as larger precursor proteins (see Kreil *et al.*, 1977; Land *et al.*, 1982).

Although careful sequence determination should produce authentic sequence, artefacts can be introduced during cloning. These reflect substantial rearrangements particularly near the 5′ terminus (relative to the sense strand) of the cloned cDNA, as has been described during studies with cloned bovine parathyroid hormone cDNA (Weaver *et al.*, 1981). Inversion at the 5′ terminus, revealed by sequence comparison with a genomic clone, has also been reported for a rat insulin cDNA insert (Chan *et al.*, 1979). Consequently care should be taken in the interpretation of DNA sequence data, particularly if it purports to encode novel peptide sequence.

The analysis of DNA sequence data by "eye" is tedious, time-consuming, subject to human error and unnecessary, since computers are now widely used and widely available for the storage, analysis, and comparison of DNA sequences. Numerous computer programmes have been assembled for this purpose. These include integrated sets of programmes together with updated libraries of sequence that are accommodated in a few large computer systems, and small sets of specialised programmes suitable for use on microcomputer systems already available in many laboratories (see: Applications of computers to research on nucleic acids, Nucleic Acids Res. (1982) *10*, No. 1). Access to such systems is now essential for the rapid and accurate processing of DNA sequence data.

V Expression of sequences encoding peptide hormones in prokaryotic and eukaryotic host-vector systems

As outlined in Fig. 1, the synthesis and post-translational processing/modification of peptide hormones can involve a series of events of differing complexity within the secretory pathway, starting with the interaction of the nascent "signal" peptide with the membrane environment of the endoplasmic reticulum, and resulting ultimately in the secretion of processed biologically active peptide(s) by the endocrine gland of origin. Consequently where the production of a particular peptide hormone using recombinant techniques is envisaged, careful consideration should be given to the manner in which hormones are synthesised *in vivo*, and in particular the importance of post-translational modifications such as proteolytic cleavage, formation of disulphide bonds for orientation of multi-chain proteins, glycosylation, acetylation, amidation, and other modifications required for biological activity of the final product. Others in this series have already considered at length the design and application of established prokaryote and eukaryote based expression vectors (see Thompson, 1982; Brammar, 1982; Rigby, 1982; Harris, 1983). We will not therefore consider in depth the design of the vectors or choice of host since many of the published systems will soon be superseded by a generation of more sophisticated vectors. For instance, host organisms more suited to large scale industrial fementations than the laboratory models are being developed, as are systems capable of performing specialised post-translational modifications, and systems designed not to maximise peptide hormone gene expression but to investigate the factors required for the regulation of their expression *in vivo*. Instead we will briefly consider the strategies which have been employed thus far to minimise the requirement for post-translational processing in the production of biologically active peptides in prokaryotes, and the future potential of eukaryote based host-vector systems.

A The synthesis of peptide hormones by prokaryotic host-vector systems

Two somewhat overlapping approaches have been used for expression studies in prokaryotic systems. The first involves the complete chemical synthesis of a gene based on the known amino-acid sequence and careful use of the genetic code. Using this approach, small peptide hormones normally synthesised as high molecular weight precursor proteins, may be synthesised without flanking peptides, to avoid the requirement for subsequent proteolytic processing events, associated with the secretory pathway *in vivo*. The second utilizes the cDNA clone. This route, whilst reducing the amount of chemical synthesis required, has the disadvantage that unless fortuitous restriction sites are available, the cDNA often encodes a precursor protein which must then be processed, either *in vivo* by the host or alternatively *in vitro* after partial purification of the peptide hormone sequence from host proteins.

1 Elevated expression of small peptide hormones as hybrid proteins

Somatostatin, a tetradecapeptide hormone known to have a role in the modulation of the secretion of certain peptide hormones and possibly also as a neurotransmittor, was the first eukaryotic polypeptide to be synthesised in bacteria. Although it is now known that two distinct somatostatins occur and that both are synthesised as high molecular weight precursor proteins as determined by cloning and sequencing of the cDNA encoding them (Hobart *et al.*, 1980), the strategy used to obtain somatostatin expression did not utilise a cloned cDNA sequence. Instead a synthetic gene was constructed from a number of overlapping oligodeoxynucleotide fragments (see Fig. 10), containing information for 15 amino acids (14 amino acids of somatostatin plus an NH_2-terminal methionine) and a stop codon. The construction took account of the degeneracy of the genetic code, using where possible *E. coli*-preferred codons, and was designed to avoid the formation of secondary structural features

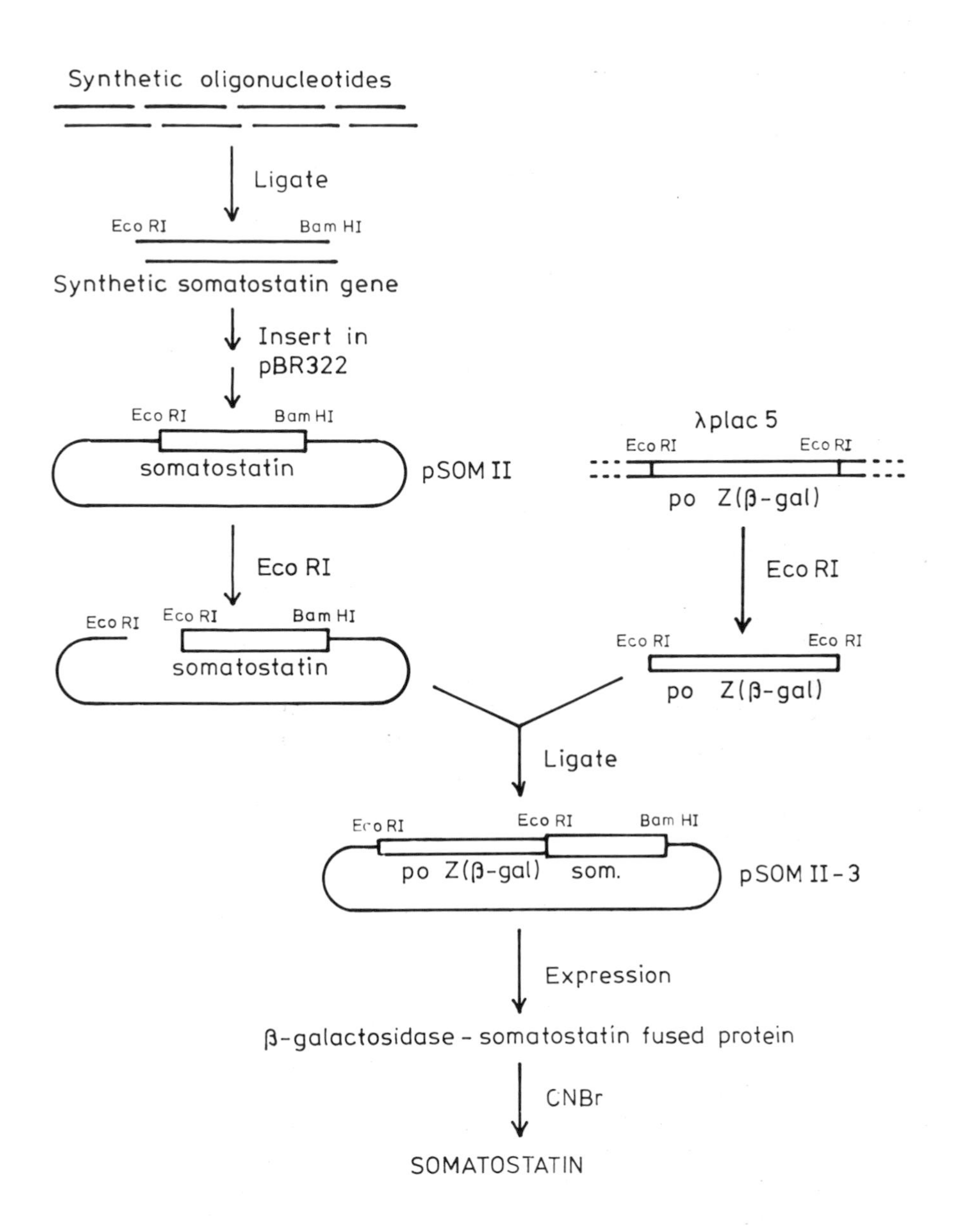

Figure 10 Bacterial expression of a cloned synthetic somatostatin gene. A recombinant plasmid, pSOMII-3 (Itakura *et al.*, 1977), was constructed which contained a synthetic somatostatin gene inserted adjacent to a cloned fragment of the *E. coli lac* operon (containing the promoter, operator and part of the β-galactosidase gene). Subsequent expression produced a β-galactosidase-somatostatin fused protein which could be cleaved with cyanogen bromide to release the somatostatin.

which might adversely affect expression. The construction also incorporated cohesive single-stranded termini for the restriction endonucleases *Eco*RI and *Bam*HI, to facilitate subsequent insertion into plasmid DNA (see Itakura *et al.*, 1977).

Using the strategy outlined in Fig. 10, the synthetic somatostatin gene was inserted along with a large fragment of the *E. coli lac* operon (containing the promoter, operator and part of the β-galactosidase gene) within the plasmid pBR332, in such a way that the β-galactosidase and somatostatin sequences were in phase with each other, and in the same orientation, resulting in the synthesis of a fusion protein, β-galactosidase-Met-somatostatin. The advantages of this construction were several fold. First, somatostatin could be recovered from the fusion protein product by treatment *in vitro* with cyanogen bromide, an efficient and specific method of cleaving polypeptides at methionine residues. Secondly, the synthetic gene was under the control of the *lac* operon, so that the synthesis of the fusion protein could be regulated by IPTG, an inducer capable of elevating β-galactosidase gene expression so that the fusion protein represented 3% of cellular protein. Thirdly, and perhaps at the time unexpectedly, apparent levels of expression of somatostatin were considerably higher when synthesised via a large fusion protein, as opposed to an alternative construction where the somatostatin gene was inserted just downstream of the *lac* operator. The latter resulted in the synthesis of a small unstable fusion protein, a reflection of rapid degradation of small peptides by *E. coli*.

An identical strategy (with some variation in the actual *lac* operon fragment used) has been employed for the expression of other small peptides including thymosin α_1 (Wetzel *et al.*, 1980), Leu-enkephalin (Shemyakin *et al.*, 1980) and α-neo-endorphin (Tanaka *et al.*, 1982). In all instances the peptide product was extensively characterised, using chemical and immunological techniques, and with the exception of the Leu-enkephalin, evidence for biological activity of the bacterially synthesised peptide was obtained. Levels of expression varied from $2–3 \times 10^4$ molecules per cell (somatostatin) to 5×10^5 molecules per cell (α-neo-endorphin).

Chemically synthesised "genes" and subsequent inducible expression of hybrid proteins has also been used for the construction of human insulin. In this instance, separate "genes" for the A and B chains were chemically synthesised (the B chain in two parts) each encoding an additional NH_2-terminal methionine, and bearing single-stranded cohesive termini to facilitate their cloning in pBR322 (see Fig. 11; also Crea *et al.*, 1978). Separate expressing plasmids were then constructed containing either the A or B chain "gene" linked in the correct orientation and phase to a fragment of the *lac* operon.

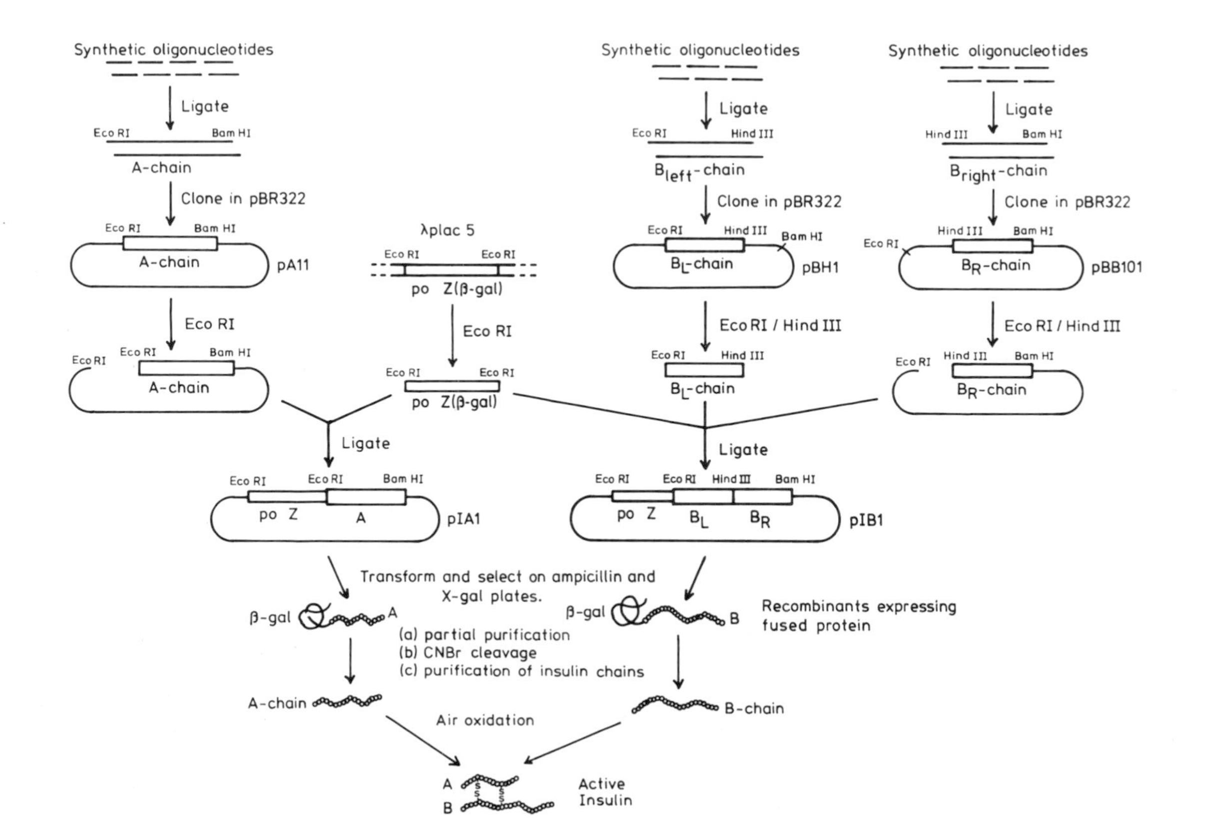

Figure 11 Bacterial expression of cloned synthetic insulin sequences. Separate recombinant plasmids, pIA1 and pIB1, containing synthetic DNA sequences encoding the human insulin A and B chains were constructed as shown (Goeddel *et al.*, 1979b). In both cases the insulin DNA was inserted adjacent to a fragment of the *E. coli lac* operon (c.f. Fig. 10) resulting in the expression of fused β-galactosidase-insulin hybrid proteins. These were subsequently cleaved with cyanogen bromide and purified, and used to assemble active insulin.

Analysis of selected colonies (Goeddel *et al.*, 1979b) showed that up to 20% of the total cellular protein was in the form of either the β-galactosidase-insulin A or B chain hybrid. The individual chains were then released using cyanogen bromide, purified and the S-sulphonated derivatives of each chain then mixed, giving rise to a 10–15% yield of human insulin as judged by a radioimmunoassay based on the reconstitution of complete insulin from separate chains.

The construction of fusion proteins whereby peptide release is dependent on cyanogen bromide cleavage is obviously limited to those peptide hormones which do not contain methionine residues. Other constructions are however possible, via selective cleavage using proteolytic enzymes after chemical modification of certain amino-acid residues to prevent unwanted proteolytic cleavage. An elegant example has been described by Shine *et al.* (1980) for the expression and release of biologically active β-endorphin. In this instance "gene" synthesis was not employed; instead a restriction fragment from cloned proopiocortin cDNA encoding mouse β-MSH and β-endorphin was recloned after suitable enzymic modifications and the use of linkers, to give a β-galactosidase-β-MSH/β-endorphin hybrid protein. To release β-endorphin, which contains no arginine residues, internal lysine residues were protected from tryptic digestion by reaction with citraconic anhydride. Digestion with trypsin then resulted in cleavage at the arginine residue immediately preceding the endorphin sequence, and also at a number of other sites within the remainder of the fusion protein. Citraconic groups were then removed by mild acid hydrolysis and after partial purification the β-endorphin was shown to have opiate activity as judged by a number of criteria. A similar approach using trypsin has also been used to release β-urogastrone (53 AAs) from a fusion product linked via a Lys-Lys sequence to a portion of the *E. coli trp* E gene (Smith *et al.*, 1982).

The constructions described side-step the issue of post-translational maturation by specific proteolytic processing enzymes, but emphasize the problems of post-translational modification in two respects. First, thymosin α_1 as isolated from the thymus has an acetylated NH_2-terminal serine. Fortuitously the absence of this modification in the bacterial product — N^{α}-desacetyl-thymosin α_1 — does not appear to affect biological activity. However, it seems unlikely that this will prove to be a universal phenomenon; consequently if prokaryotes cannot perform the necessary modifications, either these must be carried out *in vitro*, or alternative host systems capable of post-translational modification must be developed. Secondly, in spite of the successful production of individual insulin A and B chains, the final steps in the synthesis of biologically active insulin, the formation of correctly orientated chains, results in low yields. Recent work points to a possible means of averting this problem (see Wetzel *et al.*, 1981), via the synthesis of a β-galactosidase-Met-proinsulin-like protein. In this construction, the plasmid contains both the human insulin A and B chain synthetic DNA "genes", linked by an 18 nucleotide sequence which encodes a "mini-C" peptide of 6 amino-acid residues (Arg-Arg-Gly-Ser-Lys-Arg) which include the dibasic amino acids required for proteolytic cleavage. The gene analogue, when inserted adjacent to an *E. coli lac* operon fragment, resulted in the expression of a proinsulin-like protein which could be separated from β-galactosidase by cyanogen bromide cleavage. When the prohormone was subjected to conditions favouring disulphide interchange a product resembling the correctly folded mini-C proinsulin analogue could be identified, though it has yet to be established whether the correct disulphide bonds were formed, or whether the analogue could be converted to an active form, using an *in vitro* proinsulin-processing system composed of trypsin and carboxypeptidase B (Kemmler *et al.*, 1971). However, where alignment and linkage of non-identical peptide chains is required, synthesis of the prohormone followed by cleavage *in vitro* may prove a more efficient route, than the separate synthesis of the individual peptides followed by alignment *in vitro*.

2 *Expression of large peptide hormone and prohormone sequences*

Strategies employed for the expression of large peptide hormone genes differ from those used to express small peptide hormones. The hybrid protein construction, although used in early experiments to study the feasibility of the production of eukaryotic cell proteins in bacteria (Villa-Komaroff *et al.*, 1978), suffers from the inherent drawback that the larger the peptide hormone, the less likely it is to be recovered from a fusion protein, except under fortuitous circumstances where methionine residues are absent from the polypeptide, or blocking groups may be used. Thus the larger the protein the less practical the hybrid strategy. Moreover, the larger the protein the less practical it becomes to synthesise a complete gene chemically. Both problems may be overcome using a different strategy which avoids the fusion protein, and uses a combination of chemically synthesised oligonucleotides and cloned cDNA sequences.

Such a strategy has been used in the development of an expression system for the production of human growth hormone, in which the gene (part cDNA and part chemically synthesised DNA) was linked to a bacterial promoter resulting in the expression of a Met-growth

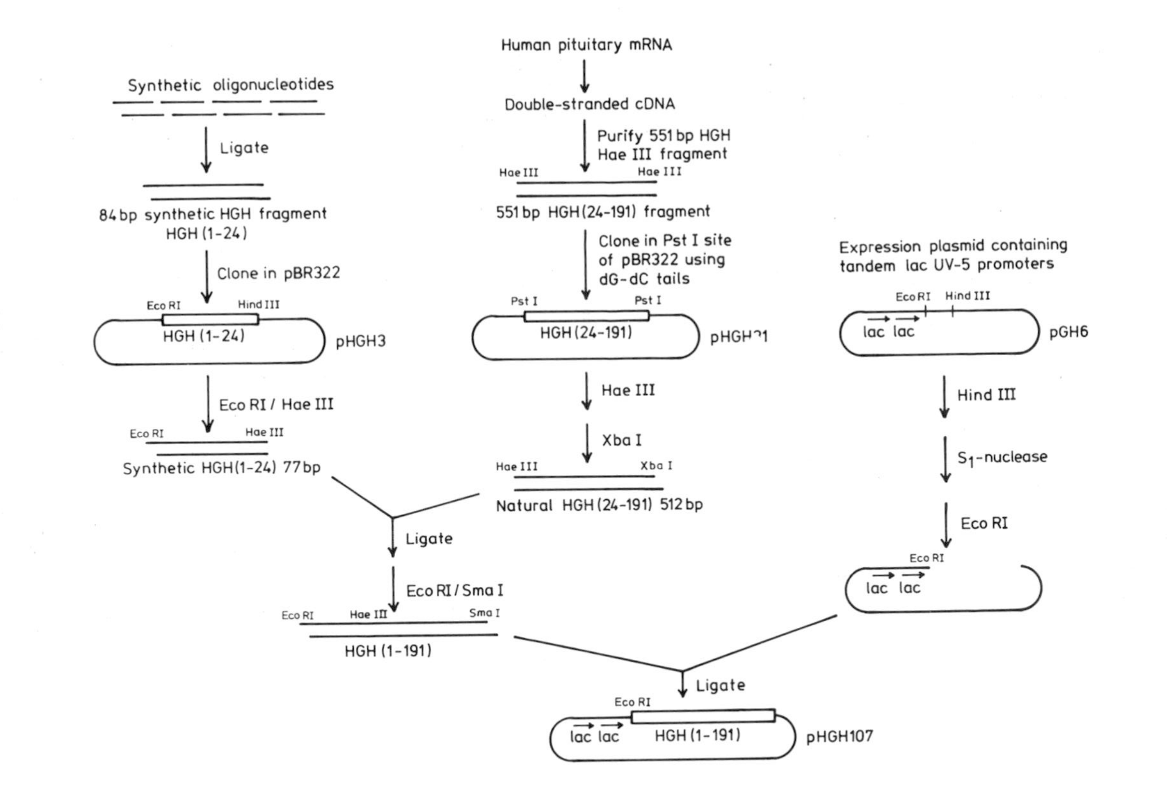

Figure 12 Bacterial expression of a cloned human growth hormone cDNA sequence. A recombinant plasmid, pHGH107, was constructed which contained a human-growth hormone "gene" (part cDNA and part chemically synthesised DNA) linked to tandem *lac* promoters (Goeddel *et al.*, 1979a). Extracts of bacteria carrying this plasmid contained biologically active Met-HGH (for further details see text).

hormone polypeptide (see Fig. 12). The construction used a 551 bp *Hae*III restriction fragment, isolated from a plasmid containing growth hormone cDNA and shown to encode amino acids 24 to 191 of human growth hormone. This fragment, after a series of modification steps, was ligated in the correct orientation and phase to a chemically synthesised adapter fragment which encoded an NH_2-terminal methionine residue followed by amino acids 1—24 of the mature human growth hormone, but did not encode a signal peptide. The resulting DNA which encoded Met-HGH_{1-191} was then inserted downstream from 2 *lac* promotors placed in tandem. Extracts of bacteria carrying this plasmid contained growth hormone as judged by radioimmunoassay, whilst expression was shown to be under *lac* operon control (see Goeddel *et al.*, 1979a). The growth hormone produced in this manner, unlike the smaller peptides, appeared to be relatively stable, and was present at 2×10^5 molecules per cell in log-phase cultures grown in rich media. Human growth hormone (Met-HGH) synthesised in this manner proved to have biological activity, in spite of having retained the initiating methionine (Olson *et al.*, 1981).

The strategy described above creates a potentially unlimited source of a molecule with many characteristics of human growth hormone, which like the fusion protein constructions described in the previous section, accumulates in the bacterial cell, and requires extensive purification from bacterial proteins. An alternative strategy results in the secretion of the product by the bacteria (*E. coli*), or at least transportation across a membrane and sequestration within the periplasmic space making such proteins easier to detect and purify, a commercial consideration. However, there is an additional fascination of academic interest, inherent in this alternative strategy.

The signals necessary for the access of a protein to the secretory pathway lie within the primary translation product. Bacterial secretory proteins in common with most secreted proteins in higher organisms are synthesised as pre-proteins with nascent NH_2-terminal signal peptide extensions (Bedouelle *et al.*, 1980; Emr *et al.*, 1980; see section II.C). Thus when rat proinsulin cDNA was inserted using G/C tailing into the *Pst*I site of the plasmid pBR322, a site which lies within the plasmid gene encoding the secreted protein β-lactamase, a number of colonies containing recombinant plasmids retained resistance to ampicillin. Subsequent analysis showed that a β-lactamase_{1-182} $(\text{Gly})_6$-proinsulin hybrid protein which retained β-lactamase activity was being synthesised (see Villa-Komaroff *et al.*, 1978), since the proinsulin cDNA and the connecting G/C tails encoding $(\text{Gly})_6$ were in the correct orientation and in phase with the β-lactamase gene. More important, the fusion protein could be recovered from the periplasmic space, indicating that in spite of the hybrid nature of the protein, the "signal" peptide of β-lactamase retained its function. Subsequent experiments using preproinsulin or proinsulin cDNA cloned into plasmids constructed with a *Pst*I site in or near the β-lactamase signal sequence (Talmadge and Gilbert, 1980) showed that not only was preproinsulin processed to proinsulin, but that a eukaryotic signal sequence was sufficient to transport insulin antigen to the periplasmic space in *E. coli* (see also Talmadge *et al.*, 1980a, b; 1981). Although high levels of expression were not sought, the strategy provides a means of synthesising a eukaryotic secretory protein or prohormone which is neither fused to bacterial protein nor contains an additional NH_2-terminal methionine residue.

B Synthesis of peptide hormones by eukaryotic host-vector systems

Early work using prokaryotic expression systems emphasised the desire to synthesise in quantity biologically active peptides of clinical, and therefore potentially commercial, significance. The outcome, though highlighting the problems of intracellular proteolysis, and the absence in bacteria of specialised enzymes required for the post-translational modification/processing of prohormones, demonstrates that in theory at least, any peptide can be produced using recombinant techniques in essentially unlimited quantity.

A number of systems have been successfully utilised for the reintroduction and expression of mammalian gene sequences in cultured eukaryotic cells. All have proved to be of considerable academic interest, since they have provided the means with which to investigate the factors and regulatory gene sequences required to modulate the expression of structural genes, and in a more general sense, the intracellular mechanisms involved in the synthesis, post-translational modification and secretion of mammalian proteins (see Rigby, 1982). Published studies to date on the expression of peptide hormones have been limited to the expression of the rat preproinsulin gene, and the rat and human growth hormone genes using either direct cotransformation with the Herpes Simplex virus thymidine kinase gene into tk^--cells (Robins *et al.*, 1982) or viral vectors based on SV40 (Gruss and Khoury, 1981; Pavlakis *et al.*, 1981), retrovirus (Doehmer *et al.*, 1982) and papilloma virus DNA (Sarver *et al.*, 1981). Techniques have relied entirely on previously cloned genomic DNA, avoiding the costly requirement of oligodeoxynucleotide synthesis and the intricate constructions necessary for prokaryotic expression strategies.

The utilisation of SV40 vectors, though limited in host range, with associated problems of host cell death, and the disadvantage in terms of regulatory studies that transcription of inserted sequences occurred via SV40 promoters, demonstrated perhaps not unexpectedly in the light of earlier oocyte mRNA microinjection studies (see section III.C.1) that whereas mammalian cells will express an introduced rat preproinsulin gene (Gruss and Khoury, 1981; Gruss *et al.*, 1981) the secreted product was proinsulin. Thus mammalian cell lines in common with bacterial systems, lack specialised proteolytic processing enzymes, but will sequester within the secretory pathway and secrete the synthesised protein. Predictably, similar independent studies based on an SV40 vector, using human growth hormone genes (Pavlakis *et al.*, 1981), demonstrated that processed growth hormone (i.e. minus the signal peptide) could be synthesised and secreted. The studies on the expression of human growth hormone have proved of interest for reasons other than the now recognised use of deletion mutants to explore intracellular processing mechanisms (see Gruss *et al.*, 1981). The human genome contains a family of at least seven growth hormone related genes (Moore *et al.*, 1982), yet although studies at the protein level support the view of a complex growth hormone family, possibly a reflection of variable post-translational modifications, only 2 growth hormone protein variants have been purified from pituitary extracts. One, present at relatively low levels in the normal pituitary, lacks amino-acid residues 32—46 of the major form (see Lewis *et al.*, 1980) and probably arises via an alternative splicing pathway of a single gene transcript as opposed to representing transcription of a separate gene (see Moore *et al.*, 1982). The question therefore arises as to whether the remaining 6 related genes are pseudogenes (which cannot be expressed due to the presence of mutations), or are potentially expressible.

The insertion of such "unknown" gene sequences into a eukaryotic expression system permits such problems to be investigated, and moreover, if expression occurs, provides sufficient secreted protein ($50\,\mu g/2 \times 10^7$ cells/d) to allow preliminary analyses for bioactivity. Using this approach Pavlakis and co-workers (1981) were able to demonstrate that the human growth hormone like gene (hGH2) which contained 14 point mutations, all which would be expected to lead to amino acid substitutions when compared with the major form (hGH1), directed the synthesis and secretion in monkey cells of a growth hormone-like protein. This differed from authentic human growth hormone in its behaviour on isoelectric focusing, and in its low immunoreactivity with anti-hGH serum, yet bound efficiently in a receptor assay. Whether this growth hormone variant gene is expressed *in vivo*, and therefore is of physiological and possibly clinical significance remains to be established. However, the observation that it is potentially expressible should encourage further studies.

Expression systems based on cotransformation using the thymidine kinase gene as selectable marker provide a means of obtaining expression of an introduced gene without cell death, but with the associated disadvantage, that the transcriptional activity of a single cloned gene varies between selected transformants, reflecting rearrangements and/or integration into different regions of the host genome (see Scangos and Ruddle, 1981). The analysis of expressed genes using this system can therefore prove problematical in instances where a cell-line contains multiple copies or fragments of a cloned gene. Some of these may be expressed at low levels, some not at all, whilst in some instances the expression may be modulated by the appropriate hormonal milieu. For example the enhancement of human growth hormone synthesis and secretion by glucocorticoids occurs in some but not all cell lines established using this procedure (Robins *et al.*, 1982).

Expression of the rat growth hormone gene has also been reported using a retroviral DNA vector. Here the ability of the viral DNA to transform recipient cells provides the selection criteria (Doehmer *et al.*, 1982), effectively widening the range of potential host cell lines, as tk^- mutant cell lines are no longer required for selection (see Rigby, 1982). The expression of a rat growth hormone gene in this system proved to be modulated by glucocorticoids, and furthermore, significant amounts of growth hormone (150 ng/ml/72 h) were secreted as judged by immunoreactivity and column chromatography. Examination of the transformed cell lines revealed not only the presence of between 3 and 20 integrated gene copies, but also subgenomic fragments, demonstrating that in common with the tk system not all integrations contained intact genomic sequences. Moreover, although those cell lines analysed secreted processed rat growth hormone as judged by immunoprecipitation and gel filtration, the mRNA size proved larger by 250 ntd than its rat pituitary counterpart, suggesting that retroviral vectors may not prove suitable systems with which to investigate transcriptional and post-transcriptional mechanisms, where it is intended to relate experimental observation using "introduced" genes with "normal" cellular processing events.

The development of bovine papilloma virus (BPV) as a viral vector significantly extends the potential of eukaryotic expression vectors. The unique feature of this DNA virus lies in its ability to propagate as an extrachromosomal, multiple copy plasmid (Law *et al.*, 1981;

Moar *et al.*, 1981), and to induce morphological transformation in established cell-lines, in particular mouse cells. BPV DNA and a subgenomic fragment (69% of genome) have been cloned in pBR322, and both shown to transform susceptible mouse cells, provided that inhibitory pBR322 sequences had been separated from BPV DNA by restriction endonuclease digestion (Lowy *et al.*, 1980) before transformation. Experiments using the rat preproinsulin gene demonstrated the potential of BPV as an expression vector. It could be shown that recombinants transformed mouse cells, and replicated in an extrachromosomal state. The introduced gene was transcribed from its own promoter, and the resulting transcripts processed to preproinsulin mRNA in an apparently identical manner to that observed *in vivo*. The mRNA was translated, and proinsulin, as judged by radioimmunoassay and gel electrophoresis, secreted in measurable amounts (Sarver *et al.*, 1981). The regulation of expression of human β-interferon by poly(I). poly(C) has also been demonstrated in mouse cells using the BPV vector system (Zinn *et al.*, 1982). However, in these experiments correct initiation of β-interferon transcription was not always observed. Moreover, the extrachromosomal recombinant DNA often contained foreign DNA sequences other than the interferon gene between the ends of the linear transforming DNA (see also Law *et al.*, 1981).

The development of a new BPV/pBR322-derived vector that progagates as an extrachromosomal element in both mouse and bacterial cells and does not inhibit mouse cell transformation has improved this system considerably. This construction has the advantage that all transformed mouse cell lines should contain identical extrachromosomal elements, since cloned circular DNA as opposed to linear DNA is introduced, reducing the frequency of rearrangements and the presence of extraneous DNA. Experimental analysis of the "shuttle" vector has shown that unrearranged BPV-β-globin plasmids can be recovered from transformed mouse cells by bacterial transformation using low molecular weight DNA. Moreover, the recovered DNA induced focus formation when reintroduced into mouse cells (see DiMaio *et al.*, 1982).

Mammalian expression systems are of increasing importance. The general availability of well characterised stable expression systems such as the "BPV shuttle" system described above should encourage those working with interesting biological systems, but short on specialist expertise, to branch out into pastures new — the potential is enormous. Stable expression systems in combination with *in vitro* mutagenesis (see Lathe, Lecocq and Everett, pp. 1—56) and deletion mutants permits the examination not just of upstream regulatory gene sequences and splicing mechanisms, but also nucleotide sequences specifying regions of structural significance, related for instance to enzyme structure and function, mechanisms of membrane insertion, and amino-acid sequence or secondary structure necessary for post-translational processing events (see for example Inouye *et al.*, 1982; Lomedico and McAndrew, 1982; Wasylyk *et al.*, 1980; also Rigby, 1982). Above all, *in vitro* mutagenesis in combination with a stable expression system provides a rapid means of producing hormone analogues, some of which may prove more potent than their natural "ancestors" synthesised and secreted *in vivo*.

VI Future perspectives

What has been learnt from cloning peptide hormone cDNA sequences? How can the knowledge gained be best utilized and extended in future years?

It is now clear that the construction of representative cDNA libraries presents no real obstacle, though the screening of these libraries for specific recombinants may not be a trivial matter, particularly where the sequence of interest is of low abundance, and/or selection relies on a bioassay. However, well tried strategies now exist which should permit the cloning of any mRNA species encoding a biologically active molecule, whether peptide hormone or growth factor. Significant advances in our understanding of the structure and expression of peptide hormones have been made, simply through nucleotide sequence analysis of cloned cDNA sequences and the availability of sequence specific hybridisation probes. This has perhaps been most revealing when considered in terms of the small peptide hormones.

Recombinant techniques have established that all so far examined, are synthesised as high molecular weight precursor polyproteins, either as a precursor protein containing a number of biologically active peptides or multiple copies of the same peptide (see Fig. 13). In some instances this has defined the ordering of peptides previously suspected to be synthesized within a single polyprotein (Land *et al.*, 1982), but in others has identified novel or "cryptic" peptides of unknown biological significance. Chemical synthesis of "predicted" peptides opens up important research areas of fundamental and clinical relevance. Synthetic peptides, either the whole peptide, or those regions of the peptide most likely to reside on the outside of larger molecules and so prove most antigenic (see Chou and Fasman, 1978; Hopp and Woods, 1981) may be used to raise polyclonal or monoclonal antibodies. These can then be used to monitor circulating peptide levels in health and disease, to determine tissue

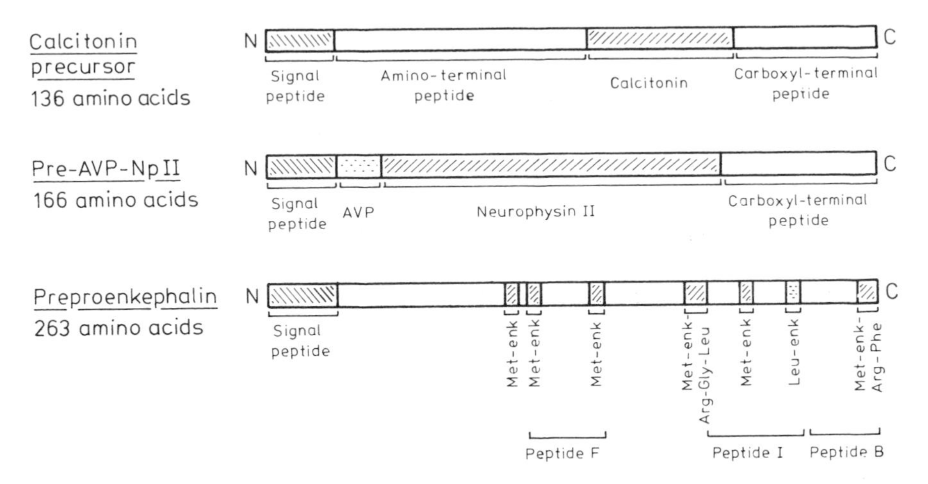

Figure 13 Structure of some characterised peptide hormone polyproteins deduced from DNA sequence analysis of cloned mRNA species. Rat prepro-calcitonin precursor — from Jacobs *et al.* (1981). Bovine prearginine vasopressin-neurophysin II precursor — from Land *et al.* (1982). Bovine preproenkephalin precursor — from Noda *et al.* (1982).

distribution, and to investigate intracellular post-translational processing events. The potential hormonal or neurotransmitter activity of synthetic peptides can also be investigated. These approaches have been used to demonstrate: (a) that the C-terminal flanking peptide predicted by nucleotide sequence analysis of the rat calcitonin mRNA (Fig. 13) is present in rat thyroid tissue, and is secreted by monolayer cultures of rat medullary thyroid carcinoma (Birnbaum *et al.*, 1982), and (b) that the equivalent human carboxyl terminal peptide (see Fig. 9) circulates in man, has biological activity in rats, and may therefore represent a new calcium-lowering hormone in man, and moreover, may also play a role as a neuromodulator in the regulation of pituitary function (MacIntyre *et al.*, 1982). Others have reported a physiological role for an NH_2-terminal fragment of human proopiocortin precursor (Pedersen and Brownie, 1980; Pedersen *et al.*, 1980; Seidah *et al.*, 1981). The potential of this approach is endless. Numerous small biologically active peptides of varying tissue distribution and specificity have been identified. All presumably are synthesised as high molecular weight precursor polyproteins. Recombinant techniques provide the means of identifying flanking peptides, which may then be analysed for biological activity and function, and ultimately evaluated for their diagnostic and therapeutic potential.

Cloned peptide hormone cDNA sequences may be used as sequence specific hybridisation probes to investigate genomic organisation, in a preliminary manner via DNA blotting protocols, or in detail after selection of cloned genomic sequences from existing genomic libraries (see Dahl *et al.*, 1981). Studies such as these have resulted in the identification and characterisation of multigene families, a mechanism by which a diversity of peptide hormones can be generated (Blundell and Humbel, 1980; Moore *et al.*, 1982) and provide insight into the evolution of these genes (see Jeffreys, 1981). In addition, knowledge of "normal" gene organisation provides a molecular handle for the study of the molecular basis of genetic diseases linked to abnormal peptide hormone gene expression. For example DNA blotting has been used to demonstrate that a familial disorder (IGHD, type A) which specifically affects the synthesis of human growth hormone is caused by deletion of the HGH-1 gene (Phillips *et al.*, 1981). Thus in common with the now well characterised haemoglobinopathies (see Little, 1981) access to a cDNA hybridisation probe permits direct correlations to be made between a single peptide hormone gene defect and a clinical phenotype.

Cloned peptide hormone cDNA sequences may also be used as sequence specific hybridisation probes to locate, using *in situ* hybridisation, those areas of tissue sections which contain specific mRNA populations (see Hudson *et al.*, 1981b). The relative abundance of a given mRNA within RNA populations and the effect of hormones on mRNA biosynthesis can be determined rapidly, on multiple samples, and with considerable accuracy, using dot hybridisation procedures (see Robins *et al.*, 1982; White and Bancroft, 1982). Normal and abnormal mRNA processing pathways can be evaluated using RNA blotting techniques, by comparison of mRNA biosynthesis in different endocrine tissues and tumours. Studies such as these have provided evidence that a proopiocortin-like mRNA is present not just in rat pituitary, but in measureable amounts in other brain tissues, such as the hypothalamus, amygdala and cerebral cortex (Civelli *et al.*, 1982). Technically simple experiments, but of importance, since they provide evidence that proopiocortin peptide cleavage products previously identified in brain tissues by immunochemical techniques are synthesised in different regions of the brain, and are not of pituitary origin. A similar approach has been used to demonstrate the expression of the prolactin gene in human placental tissue (Clements *et al.*, 1982).

Differences in size and relative abundance of calcitonin-like mRNA species as determined using RNA blotting techniques form the basis of a series of elegant experiments which have recently shown that in the rat, apparently tissue specific RNA splicing of a single gene transcript results in the generation of mRNA species encoding different polypeptide products (see Fig. 14). Thus in the thyroid,

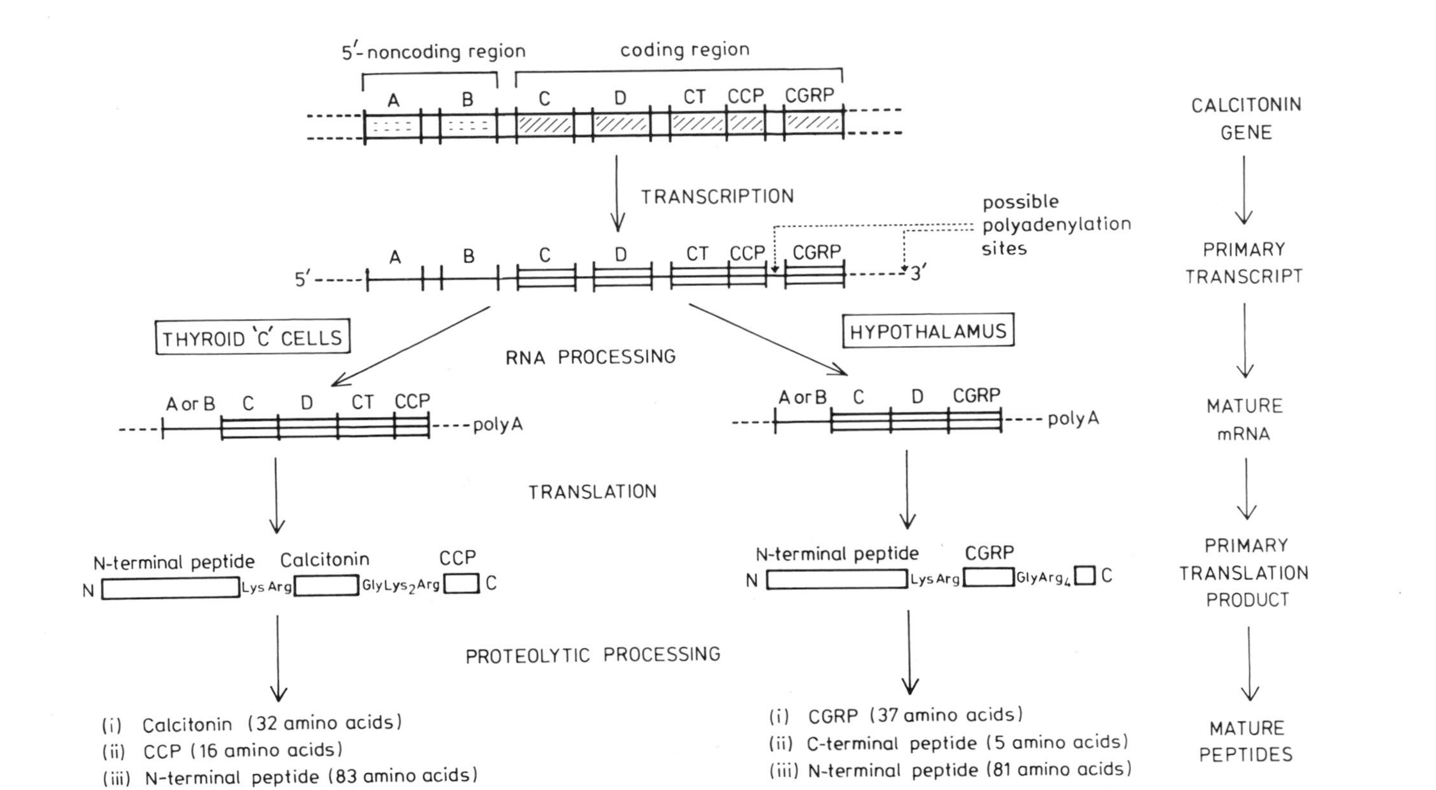

Figure 14 Tissue specific RNA processing of the rat calcitonin gene primary transcript. Adapted from Amara *et al.* (1982).

calcitonin gene expression results in the production of an mRNA encoding the calcitonin polyprotein, yet in the hypothalamus, the mRNA encodes a polyprotein containing a calcitonin gene-related peptide of as yet undetermined function (Amara *et al.*, 1982). These observations are of fundamental importance since they provide an example of a way in which tissue specific RNA processing mechanisms can be utilized to increase the diversity of neuroendocrine gene expression, in constrast to tissue specific post-translational processing mechanisms which result in a diversity of peptide products generated from a single polyprotein (see Section II.C, also Herbert and Uhler, 1982).

The experimental objectives outlined above require only limited expertise in terms of molecular techniques. Blotting technology is well established, enzyme kits for radiolabelling plasmid DNA can be purchased "off the shelf", and "predicted" peptides can be custom synthesised. In fact experimental design, and application, moves to some extent from the sphere of the molecular biologist back to the clinician, the endocrinologist, and the protein chemist, whose specialist knowledge and laboratories are better suited to perform much of the experimentation. These are the immediate spin-offs of a successful collaborative exercise. What might be the long-term objectives?

Several complementary areas of research, of commercial, academic and ultimately clinical (or for that matter agricultural) importance come to mind. Repeatedly throughout this article we have considered the problems associated with the inability of cell-free systems, and prokaryotic and eukaryotic expression systems to perform tissue specific post-translational modification and processing events. The synthesis of prohormones, polyproteins, or inappropriately modified or unmodified proteins by expression systems may not provide an unlimited source of biologically active peptides, but does provide a source of substrate for the purification and characterisation of apparently tissue specific processing enzymes, known to reside within the secretory pathway of endocrine cells. The isolation of cDNA or genomic sequences specifying these enzymes will permit a number of avenues to be explored.

Initially it will be necessary to determine whether the expression of these genes in eukaryotic host-vector systems results in the synthesis of enzymes which are then integrated into the appropriate organelle(s) within the secretory pathway and retain their required specificity. Should this prove to be the case — and this would provide in itself the basis of fascinating experiments into the intracellular mechanisms involved in the segregation of enzymes within intracellular organelles — then it seems reasonable to predict that future expression systems may well incorporate a number of inserted gene sequences. Such systems may include the peptide hormone, prohormone or polyprotein gene, and additional genes which specify the necessary processing and modification enzymes required to ensure the secretion of fully processed and modified peptide hormone(s) into the surrounding medium, so avoiding costly purification and inefficient processing procedures *in vitro*. An additional degree of sophistication might require the incorporation of upstream regulatory sequences to ensure that the diverse genes were regulated and their synthesis amplified by different external stimuli — glucocorticoids, oestrogens, androgens, small synthetic peptide hormones (see for example Lee *et al.*, 1981). In this way host cells could be switched from synthesizing the processing/modification enzymes necessary to "prime" the secretory apparatus, over to the synthesis and secretion of the required peptide hormone(s). The limitation of this "commercial" scenario would be the capacity of the host cell secretory pathway, and the presence of the required diversity of hormone receptors within the host cell line, to permit multiple hormonal manipulations.

The mechanism by which processing enzymes regulate the coordinated synthesis of functionally related peptides from a single polyprotein in a tissue specific manner is also of importance, if we are to gain insight into the manner in which small peptides mediate complex biological responses in eukaryotes (see also Herbert and Uhler, 1982). In a similar vein, where peptide diversity is a function of tissue specific RNA processing mechanisms, how is this specificity determined?

Gene cloning and related technology has already provided new insight into the structure and in some instances function of peptide hormones, their multiple sites of synthesis, and the intracellular mechanisms involved in their synthesis and secretion. The final goal remains. How does this complex array of neurotransmitter and neuromodulator molecules, synthesized in diverse areas of the brain, determine development and behavioural patterns in animals? Recombinant DNA technology has opened the door, and has much still to offer. However, it should be remembered that it is only a collection of highly sophisticated techniques but not a science in itself. The answer to the final question ultimately lies in a coordinated approach using expertise from many disciplines.

VII Acknowledgments

We are grateful to the Wellcome Trust and Cancer Research Campaign for their generous support of a variety of projects relating to the application of recombinant DNA technology to problems of clinical relevance. We also thank Miss Fiona Whichelow for her help in preparation of this manuscript.

VIII References

Agarwal, K. L., Brunstedt, J. and Noyes, B. E. (1981). *J. Biol. Chem.* **256**, 1023–1028.

Allison, J., Hall, L., MacIntyre, I. and Craig, R. K. (1981). *Biochem. J.* **199**, 725–731.

Amara, S. G., Jonas, V., Rosenfeld, M. G., Ong, E. S. and Evans, R. M. (1982). *Nature, Lond.* **298**, 240–244.

Anderson, S., Gait, M. J., Mayol, L. and Young, I. G. (1980). *Nucl. Acids Res.* **8**, 1731–1745.

Auffray, C. and Rougeon, F. (1980). *Eur. J. Biochem.* **107**, 303–314.

Aviv, H. and Leder, P. (1972). *Proc. Natn. Acad. Sci. U.S.A.* **69**, 1408–1412.

Bahl, O. P. (1977). *Fedn. Proc.* **36**, 2119–2127.

Bahl, C. P., Wu, R., Brousseau, R., Sood, A. K., Hsiung, H. M. and Narang, S. A. (1978). *Biochem. Biophys. Res. Comm.* **81**, 695–703.

Bedouelle, H., Bassford, P. J., Fowler, A. V., Zabin, I., Beckwith, J. and Hofnung, M. (1980). *Nature, Lond.* **285**, 78–81.

Bell, G. I., Swain, W. F., Pictet, R., Cordell, B., Goodman, H. M. and Rutter, W. J. (1979). *Nature, Lond.* **282**, 525–527.

Bielinska, M. and Boime, I. (1979). *Proc. Natn. Acad. Sci. U.S.A.* **76**, 1208–1212.

Birnbaum, R. S., O'Neil, J. A., Muszynski, M., Aron, D. C. and Roos, B. A. (1982). *J. Biol. Chem.* **257**, 241–244.

Birnboim, H. C. and Doly, J. (1979). *Nucl. Acids Res.* **7**, 1513–1523.

Blackburn, P., Wilson, G. and Moore, S. (1977). *J. Biol. Chem.* **252**, 5904–5910.

Blakesley, R. W. and Wells, R. D. (1975). *Nature, Lond.* **257**, 421–422.

Blobel, G. and Dobberstein, B. (1975). *J. Cell. Biol.* **67**, 835–851.

Blundell, T. L. and Humbel, R. E. (1980). *Nature, Lond.* **287**, 781–787.

Boime, I., McWilliams, D., Szczesna, E. and Camel, M. (1976). *J. Biol. Chem.* **251**, 820–825.

Bolivar, F., Rodriguez, R. L., Greene, P. J., Betlach, M. C., Heyneker, H. L., Boyer, H. W., Crosa, J. H. and Falkow, S. (1977). *Gene* **2**, 95–113.

Bonner, W. M. and Laskey, R. A. (1974). *Eur. J. Biochem.* **46**, 83–88.

Boyer, H. W. and Roulland-Dussoix, D. (1969). *J. Mol. Biol.* **41**, 459–472.

Brammar, W. J. (1982). *In* "Genetic Engineering" (R. Williamson, ed.) vol. 3, 53–81. Academic Press, London and New York.

Broome, S. and Gilbert, W. (1978). *Proc. Natn. Acad. Sci. U.S.A.* **75**, 2746–2749.

Brownstein, M. J., Russell, J. T. and Gainer, H. (1980). *Science* **207**, 373–378.

Challberg, M. D. and Englund, P. T. (1980). *In* "Methods in Enzymology" (L. Grossman and K. Moldave, eds) vol. 65, 39–43. Academic Press, New York.

Chan, S. J., Keim, P. and Steiner, D. F. (1976). *Proc. Natn. Acad. Sci. U.S.A.* **73**, 1964–1968.

Chan, S. J., Noyes, B. E., Agarwal, K. L. and Steiner, D. F. (1979). *Proc. Natn. Acad. Sci. U.S.A.* **76**, 5036–5040.

Chirgwin, J. M., Przybyla, A. E., MacDonald, R. J. and Rutter, W. J. (1979). *Biochemistry* **18**, 5294–5299

Chou, P. Y. and Fasman, G. D. (1978). *Adv. Enzymol.* **47**, 45–148.

Christophe, D., Mercken, L., Brocas, H., Pohl, V. and Vassart, G. (1982). *Eur. J. Biochem.* **122**, 461–469.

Civelli, O., Birnberg, N. and Herbert, E. (1982). *J. Biol. Chem.* **257**, 6783–6787.

Clements, J., Whitfeld, P., Cooke, N., Healy, D., Shine, J. and Funder, J. W. (1982). Abstract. Proceedings of XII International Congress of Biochemistry Meeting, Perth, Australia. p. 132.

Colman, A., Lane, C. D., Craig, R., Boulton, A., Mohun, T. and Morser, J. (1981). *Eur. J. Biochem.* **113**, 339–348.

Colman, A. and Morser, J. (1979). *Cell* **17**, 517–526.

Comb, M., Herbert, E. and Crea, R. (1982a). *Proc. Natn. Acad. Sci. U.S.A.* **79**, 360–364.

Comb, M., Seeburg, P. H., Adelman, J., Eiden, L. and Herbert, E. (1982b). *Nature, Lond.* **295**, 663–666.

Cooke, N. E., Coit, D., Shine, J., Baxter, J. D. and Martial, J. A. (1981). *J. Biol. Chem.* **256**, 4007–4016.

Craig, R. K., Brown, P. A., Harrison, O. S., McIlreavy, D. and Campbell, P. N. (1976). *Biochem. J.* **160**, 57–74.

Craig, R. K., Hall, L. Edbrooke, M. R., Allison, J. and MacIntyre, I. (1982). *Nature, Lond.* **295**, 345–347.

Craig, R. K., Hall, L., Parker, D. and Campbell, P. N. (1981). *Biochem. J.* **194**, 989–998.

Crea, R., Kraszewski, A., Hirose, T. and Itakura, K. (1978). *Proc. Natn. Acad. Sci. U.S.A.* **75**, 5765–5769.

Curtis, R., Pereira, D. A., Hsu, J. C., Hull, S. C., Clark, J. E., Maturin, C. J., Goldschmidt, R., Moody, R., Inoue, M. and Alexander, L. (1977). *In* "Recombinant Molecules: Impact on Science and Society" (R. F. Beers and E. G. Bassett, eds) 45–56. Raven Press, New York.

Dahl, H. H., Flavell, R. A. and Grosveld, F. G. (1981). *In* "Genetic Engineering" (R. Williamson, ed) vol. 2, 49–127. Academic Press, London and New York.

Daniels-McQueen, S., McWilliams, D., Birken, S., Canfield, R., Landefeld, T. and Boime, I. (1978). *J. Biol. Chem.* **253**, 7109–7114.

Deng, G. and Wu, R. (1981). *Nucl. Acids Res.* **9**, 4173–4188.

DiMaio, D., Treisman, R. and Maniatis, T. (1982). *Proc. Natn. Acad. Sci. U.S.A.* **79**, 4030–4034.

Doehmer, J., Barinaga, M., Vale, W., Rosenfeld, M. G., Verma, I. M. and Evans, R. M. (1982). *Proc. Natn. Acad. Sci. U.S.A.* **79**, 2268–2272.

Emr, S. D., Hedgpeth, J., Clement, J.-M., Silhavy, T. J. and Hofnung, M. (1980). *Nature, Lond.* **285**, 82–85.

Fiddes, J. C. and Goodman, H. M. (1979). *Nature, Lond.* **281**, 351–356.

Fletcher, D. J., Noe, B. D., Bauer, G. E. and Quigley, J. P. (1980). *Diabetes* **29**, 593–599.

Fletcher, D. J., Quigley, J. P., Bauer, G. E. and Noe, B. D. (1981). *J. Cell Biol.* **90**, 312–322.

Ghosh, P. K., Reddy, V. B., Piatak, M., Lebowitz, P. and Weissman, S. M. (1980). *In* "Methods in Enzymology" (L. Grossman and K. Moldave, eds) vol. 65, 580–595. Academic Press, New York.

Girgis, S. I., GalanGalan, F., Arnett, T. R., Rogers, R. M., Bone, Q., Ravazzola, M. and MacIntyre, I. (1980). *J. Endocrinol.* **87**, 375–382.

Giudice, L. C. and Weintraub, B. D. (1979). *J. Biol. Chem.* **254**, 12679–12683.

Glisin, V., Crkvenjakov, R. and Byus, C. (1974). *Biochemistry* **13**, 2633–2637.

Godine, J. E., Chin, W. W. and Habener, J. F. (1980). *J. Biol. Chem.* **255**, 8780–8783.

Goeddel, D. V., Heyneker, H. L., Hozumi, T., Arentzen, R., Itakura, K., Yansura, D. G., Ross, M. J., Miozzari, G., Crea, R. and Seeburg, P. H. (1979a). *Nature, Lond.* **281**, 544–548.

Goeddel, D. V., Kleid, D. G., Bolivar, F., Heyneker, H. L., Yansura, D. G., Crea, R., Hirose, T., Kraszewski, A., Itakura, K. and Riggs, A. D. (1979b). *Proc. Natn. Acad. Sci. U.S.A.* **76**, 106–110.

Grantham, R., Gautier, C., Gouy, M., Jacobzone, M. and Mercier, R. (1981). *Nucl. Acids Res.* **9**, r43–r74.

Gray, P. W., Leung, D. W., Pennica, D., Yelverton, E., Najarian, R., Simonsen, C. C., Derynck, R., Sherwood, P. J., Wallace, D. M., Berger, S. L., Levinson, A. D. and Goeddel, D. V. (1982). *Nature, Lond.* **295**, 503–508.

Grunstein, M. and Hogness, D. S. (1975). *Proc. Natn. Acad. Sci. U.S.A.* **72**, 3961–3965.

Gruss, P., Efstratiadis, A., Karathanasis, S., Konig, M. and Khoury, G. (1981). *Proc. Natn. Acad. Sci. U.S.A.* **78**, 6091–6095.

Gruss, P. and Khoury, G. (1981). *Proc. Natn. Acad. Sci. U.S.A.* **78**, 133–137.

Gubler, U., Kilpatrick, D. L., Seeburg, P. H., Gage, L. P. and Udenfriend, S. (1981). *Proc. Natn. Acad. Sci. U.S.A.* **78**, 5485–5487.

Gubler, U., Seeburg, P., Hoffman, B. J., Gage, L. P. and Udenfriend, S. (1982). *Nature, Lond.* **295**, 206–208.

Habener, J. F. and Potts, J. T. (1978a). *New Engl. J. Med.* **299**, 580–585.

Habener, J. F. and Potts, J. T. (1978b). *New Engl. J. Med.* **299**, 635–644.

Habener, J. F., Rosenblatt, M., Kemper, B., Kronenberg, H. M., Rich, A. and Potts, J. T. (1978). *Proc. Natn. Acad. Sci. U.S.A.* **75**, 2616–2620.

Haley, J., Hudson, P., Scanlon, D., John, M., Cronk, M., Shine, J., Tregear, G. and Niall, H. (1982). *DNA* **1**, 155–162.

Hall, L., Craig, R. K. and Campbell, P. N. (1979). *Nature, Lond.* **277**, 54–56.

Hall, L., Craig, R. K., Edbrooke, M. R. and Campbell, P. N. (1982). *Nucl. Acids Res.* **10**, 3503–3515.

Hall, L., Davies, M. S. and Craig, R. K. (1981). *Nucl. Acids Res.* **9**, 65–84.

Hamlyn, P. H., Brownlee, G. G., Cheng, C. C., Gait, M. J. and Milstein, C. (1978). *Cell* **15**, 1067–1075.

Hamlyn, P. H., Gait, M. J. and Milstein, C. (1981). *Nucl. Acids Res.* **9**, 4485–4494.

Hanahan, D. and Meselson, M. (1980). *Gene* **10**, 63–67.

Harpold, M. M., Dobner, P. R., Evans, R. M. and Bancroft, F. C. (1978). *Nucl. Acids Res.* **5**, 2039–2053.

Harris, T. (1982). *In* "Genetic Engineering" (R. Williamson, ed.) vol. 4, 127–185. Academic Press, New York and London.

Heidecker, G., Messing, J. and Gronenborn, B. (1980). *Gene* **10**, 69–73.

Hendy, G. N., Kronenberg, H. M., Potts, J. T. and Rich, A. (1981). *Proc. Natn. Acad. Sci. U.S.A.* **78**, 7365–7369.

Herbert, E. (1981). *TIBS* **6**, 184–188.

Herbert, E., Budarf, M., Phillips, M., Rosa, P., Policastro, P. and Oates, E. (1980). *Ann. N.Y. Acad. Sci.* **343**, 79–93.

Herbert, E. and Uhler, M. (1982). *Cell* **30**, 1–2.

Hilz, H., Wiegers, U. and Adamietz, P. (1975). *Eur. J. Biochem.* **56**, 103–108.

Hobart, P., Crawford, R., Shen, L., Pictet, R. and Rutter, W. J. (1980). *Nature, Lond.* **288**, 137–141.

Holmes, D. S. and Quigley, M. (1981). *Anal. Biochem.* **114**, 193–197.
Hook, V. Y. H., Eiden, L. E. and Brownstein, M. J. (1982). *Nature, Lond.* **295**, 341–342.
Hopp, T. P. and Woods, K. R. (1981). *Proc. Natn. Acad. Sci. U.S.A.* **78**, 3824–3828.
Houghton, M., Eaton, M. A. W., Stewart, A. G., Smith, J. C., Doel, S. M., Catlin, G. H., Lewis, H. M., Patel, T. P., Emtage, J. S., Carey, N. H. and Porter, A. G. (1980). *Nucl. Acids Res.* **8**, 2885–2894.
Hudson, P., Haley, J., Cronk, M., Shine, J. and Niall, H. (1981a). *Nature, Lond.* **291**, 127–131.
Hudson, P., Penschow, J., Shine, J., Ryan, G., Niall, H. and Coghlan, J. (1981b). *Endocrinology* **108**, 353–356.
Hughes, J., Smith, T. W., Kosterlitz, H. W., Fothergill, L. A., Morgan, B. A. and Morris, H. R. (1975). *Nature, Lond.* **258**, 577–579.
Inouye, S., Soberon, X., Francschini, T., Nakamura, K., Itakura, K. and Inouye, M. (1982). *Proc. Natn. Acad. Sci. U.S.A.* **79**, 3438–3441.
Itakura, K., Hirose, T., Crea, R., Riggs, A. D., Heyneker, H. L., Bolivar, F. and Boyer, H. W. (1977). *Science* **198**, 1056–1063.
Itakura, K. and Riggs, A. D. (1980). *Science* **209**, 1401–1405.
Jacobs, J. W., Goodman, R. H., Chin, W. W., Dee, P. C., Habener, J. F., Bell, N. H. and Potts, J. T., Jr. (1981). *Science* **213**, 457–459.
Jacobs, J. W., Lund, P. K., Potts, J. T., Bell, N. H. and Habener, J. F. (1981). *J. Biol. Chem.* **256**, 2803–2807.
James, R., Niall, H., Kwok, S. and Bryant-Greenwood, G. (1977). *Nature, Lond.* **267**, 544–546.
Jeffreys, A. J. (1981). *In* "Genetic Engineering" (R. Williamson, ed.) vol. 2, 1–48. Academic Press, London and New York.
Kakidani, H., Funutani, Y., Takahashi, H., Noda, M., Morimoto, Y., Hirose, T., Asai, M., Inayama, S., Nakanishi, S. and Numa, S. (1982). *Nature, Lond.* **298**, 245–249.
Kemmler, W., Peterson, J. D. and Steiner, D. F. (1971). *J. Biol. Chem.* **246**, 6786–6791.
Keshet, E., Rosner, A., Bernstein, Y., Gorecki, M. and Aviv, H. (1981). *Nucl. Acids Res.* **9**, 19–30.
Kimura, S., Lewis, R. V., Stern, A. S., Rossier, J., Stein, S. and Udenfriend, S. (1980). *Proc. Natn. Acad. Sci. U.S.A.* **77**, 1681–1685.
Kourides, I. A., Vamvakopoulos, N. C. and Maniatis, G. M. (1979). *J. Biol. Chem.* **254**, 11106–11110.
Kreil, G., Suchanek, G. and Kindas-Mugge, I. (1977). *Fedn. Proc.* **36**, 3081–2086.
Krieger, D. T., Liotta, A. S., Brownstein, M. J. and Zimmerman, E. A. (1980). *Recent Prog. Horm. Res.* **36**, 277–344.
Kronenberg, H. M., McDevitt, B. E., Majzoub, J. A., Nathans, J., Sharp, P. A., Potts, J. T. and Rich, A. (1979). *Proc. Natn. Acad. Sci. U.S.A.* **76**, 4981–4985.
Land, H., Grez, M., Hauser, H., Lindenmaier, W. and Schutz, G. (1981). *Nucl. Acids Res.* **9**, 2251–2266.
Land, H., Schutz, G., Schmale, H. and Richter, D. (1982). *Nature, Lond.* **295**, 299–303.
Lane, C. D., Colman, A., Mohun, T., Morser, J., Champion, J., Kourides, I., Craig, R., Higgins, S., James, T. C., Applebaum, S. W., Ohlsson, R. I., Paucha, E., Houghton, M., Matthews, J. and Miflin, B. J. (1980). *Eur. J. Biochem.* **111**, 225–235.

Lathe, R., Lecocq, J. P. and Everett, R. R. (1982). *In* "Genetic Engineering" (R. Williamson, ed.) vol. 4, 1–56. Academic Press, New York and London.
Law, M.-F., Lowy, D. R., Dvoretzky, I. and Howley, P. M. (1981). *Proc. Natn. Acad. Sci. U.S.A.* **78**, 2727–2731.
Lee, F., Mulligan, R., Berg, P. and Ringold, G. (1981). *Nature, Lond.* **294**, 228–232.
Lewis, U. J., Singh, R. N. P., Tutwiler, G. F., Sigel, M. B. VanderLaan, E. F. and VanderLaan, W. P. (1980). *Recent Prog. Horm. Res.* **36**, 477–508.
Liarakos, C. D., Rosen, J. M. and O'Malley, B. W. (1973). *Biochemistry* **12**, 2809–2816.
Lingappa, V. R. and Blobel, G. (1980). *Recent Prog. Horm. Res.* **36**, 451–475.
Little, P. F. R. (1981). *In* "Genetic Engineering" (R. Williamson, ed.) vol. 1, 61–102. Academic Press, London and New York.
Loh, Y. P. and Gainer, H. (1982). *Proc. Natn. Acad. Sci. U.S.A.* **79**, 108–112.
Lomedico, P. T. and McAndrew, S. J. (1982). *Nature, Lond.* **299**, 221–226.
Lowy, D. R., Dvoretzky, I., Shober, R., Law, M.-F., Engel, L. and Howley, P. M. (1980). *Nature, Lond.* **287**, 72–74.
Lusis, A. J., Golde, D. W., Quon, D. H. and Lasky, L. A. (1982). *Nature, Lond.* **298**, 75–77.
MacIntyre, I., Evans, I. M. A., Hobitz, H. H. G., Joplin, G. F. and Stevenson, J. C. (1980). *Arthritis Rheumatism* **23**, 1139–1147.
MacIntyre, I., Hillyard, C. J., Murphy, P. K., Reynolds, J. J., Gaines Das, R. E., and Craig, R. K. (1982). *Nature, London.* **300**, 460–462.
Mains, R. E. and Eipper, B. A. (1980). *Ann. N.Y. Acad. Sci.* **343**, 94–110.
Mains, R. E., Eipper, B. A. and Ling, N. (1977). *Proc. Natn. Acad. Sci. U.S.A.* **74**, 3014–3018.
Martial, J. A., Hallewell, R. A., Baxter, J. D. and Goodman, H. M. (1979). *Science* **205**, 602–607.
Maxam, A. M. and Gilbert, W. (1977). *Proc. Natn. Acad. Sci. U.S.A.* **74**, 560–564.
Maxam, A. M. and Gilbert, W. (1980). *In* "Methods in Enzymology" (L. Grossman and K. Moldave, eds) vol. 65, 499–560. Academic Press, New York.
McDonell, M. W., Simon, M. N. and Studier, F. W. (1977). *J. Mol. Biol.* **110**, 119–146.
Meyer, D. I., Krause, E. and Dobberstein, B. (1982). *Nature, Lond.* **297**, 647–650.
Moar, M. H., Campo, M. S., Laird, H. and Jarrett, W. F. H. (1981). *Nature, Lond.* **293**, 749–751.
Moore, D. D., Conkling, M. A. and Goodman, H. M. (1982). *Cell* **29**, 285–286.
Nakanishi, S., Inoue, A., Kita, T., Nakamura, M., Chang, A. C. Y., Cohen, S. N. and Numa, S. (1979). *Nature, Lond.* **278**, 423–427.
Nakanishi, S., Inoue, A., Kita, T., Numa, S., Chang, A. C. Y., Cohen, S. N., Nunberg, J. and Schimke, R. T. (1978). *Proc. Natn. Acad. Sci. U.S.A.* **75**, 6021–6025.
Nakanishi, S., Inoue, A., Taii, S. and Numa, S. (1977). *FEBS Letters* **84**, 105–109.
Niall, H. D., Hogan, M. L., Sauer, R., Rosenblum, I. Y. and Greenwood, F. C. (1971). *Proc. Natn. Acad. Sci. U.S.A.* **68**, 866–869.
Nilson, J. H., Thomason, A. R., Horowitz, S., Sasavage, N. L., Blenis, J., Albers, R., Salser, W. and Rottman, F. M. (1980). *Nucl. Acids Res.* **8**, 1561–1573.
Noda, M., Furutani, Y., Takahashi, H., Toyosato, M., Hirose, T., Inayama, S., Nakanishi, S. and Numa, S. (1982). *Nature, Lond.* **295**, 202–206.

Noyes, B. E., Mevarech, M., Stein, R. and Agarwal, K. L. (1979). *Proc. Natn. Acad. Sci. U.S.A.* **76**, 1770–1774.
Ohmori, H., Tomizawa, J. and Maxam, A. M. (1978). *Nucl. Acids Res.* **5**, 1479–1485.
Okayama, H. and Berg, P. (1982). *Mol. Cell. Biol.* **2**, 161–170.
Olson, K. C., Fenno, J., Lin, N., Harkins, R. N., Snider, C., Kohr, W. H., Ross, M. J., Fodge, D., Prender, G. and Stebbing, N. (1981). *Nature, Lond.* **293**, 408–411.
Palade, G. (1975). *Science* **189**, 347–358.
Palatnik, C. M., Storti, R. V. and Jacobson, A. (1979). *J. Mol. Biol.* **128**, 371–395.
Pavlakis, G. N., Hizuka, N., Gorden, P., Seeburg, P. and Hamer, D. H. (1981). *Proc. Natn. Acad. Sci. U.S.A.* **78**, 7398–7402.
Payvar, F. and Schimke, R. T. (1979). *J. Biol. Chem.* **254**, 7637–7642.
Peacock, S. L., McIver, C. M. and Monahan, J. J. (1981). *Biochim. Biophys. Acta* **655**, 243–250.
Pedersen, R. C. and Brownie, A. C. (1980). *Proc. Natn. Acad. Sci. U.S.A.* **77**, 2239–2243.
Pedersen, R. C., Brownie, A. C. and Ling, N. (1980). *Science* **208**, 1044–1046.
Pelham, H. R. B. and Jackson, R. J. (1976). *Eur. J. Biochem.* **67**, 247–256.
Phillips, J. A., Hjelle, B. L., Seeburg, P. H. and Zachmann, M. (1981). *Proc. Natn. Acad. Sci. U.S.A.* **78**, 6372–6375.
Pierce, J. G. and Parsons, T. F. (1981). *Ann. Rev. Biochem.* **50**, 465–495.
Rapoport, T. A. (1981). *Eur. J. Biochem.* **115**, 665–669.
Retzel, E. F., Collett, M. S. and Faras, A. J. (1980). *Biochemistry* **19**, 513–518.
Rigby, P. W. J. (1982). *In* "Genetic Engineering" (R. Williamson, ed.) vol. 3, 83–141. Academic Press, London and New York.
Rinderknecht, E. and Humbel, R. E. (1978). *FEBS Letters* **89**, 283–286.
Roberts, B. E. and Paterson, B. M. (1973). *Proc. Natn. Acad. Sci. U.S.A.* **70**, 2330–2334.
Roberts, J. L. and Herbert, E. (1977a). *Proc. Natn. Acad. Sci. U.S.A.* **74**, 4826–4830.
Roberts, J. L. and Herbert, E. (1977b). *Proc. Natn. Acad. Sci. U.S.A.* **74**, 5300–5304.
Roberts, J. L., Seeburg, P. H., Shine, J., Herbert, E., Baxter, J. D. and Goodman, H. M. (1979). *Proc. Natn. Acad. Sci. U.S.A.* **76**, 2153–2157.
Robins, D. M., Paek, I., Seeburg, P. H. and Axel, R. (1982). *Cell* **29**, 623–631.
Roskam, W. G. and Rougeon, F. (1979). *Nucl. Acids Res.* **7**, 305–320.
Rossier, J., Trifaro, J. M., Lewis, R. V., Lee, R. W. H., Stern, A., Kimura, S., Stein, S. and Udenfriend, S. (1980). *Proc. Natn. Acad. Sci. U.S.A.* **77**, 6889–6891.
Rougeon, F., Kourilsky, P. and Mach, B. (1975). *Nucl. Acids Res.* **2**, 2365–2378.
Ruther, U., Koenen, M., Otto, K. and Muller-Hill, B. (1981). *Nucl. Acids Res.* **9**, 4087–4098.
Sanger, F. and Coulson, A. R. (1978). *FEBS Letters* **87**, 107–110.
Sanger, F., Nicklen, S. and Coulson, A. R. (1977). *Proc. Natn. Acad. Sci. U.S.A.* **74**, 5463–5467.
Sarver, N., Gruss, P., Law, M.-F., Khoury, G. and Howley, P. M. (1981). *In* "Developmental Biology Using Purified Genes" (D. Brown and C. R. Fox, eds) vol. 23, 547–556. ICN-UCLA Symposia on Molecular and Cellular Biology. Academic Press, New York.
Saxena, B. B. and Rathnam, P. (1976). *J. Biol. Chem.* **251**, 993–1005.

Scangos, G. and Ruddle, F. H. (1981). *Gene* **14**, 1–10.
Schmale, H., Leipold, B. and Richter, D. (1979). *FEBS Letters* **108**, 311–316.
Schmale, H. and Richter, D. (1980). *FEBS Letters* **121**, 358–362.
Seeburg, P. H., Shine, J., Martial, J. A., Baxter, J. D. and Goodman, H. M. (1977a). *Nature, Lond.* **270**, 486–494.
Seeburg, P. H., Shine, J., Martial, J. A., Ullrich, A., Baxter, J. D. and Goodman, H. M. (1977b). *Cell* **12**, 157–165.
Seidah, N. G., Rochemont, J., Hamelin, J., Lis, M. and Chretien, M. (1981). *J. Biol. Chem.* **256**, 7977–7984.
Shemyakin, M. F., Chestukhin, A. V., Dolganov, G. M., Khodkova, E. M., Monastyrskaya, G. S. and Sverdlov, E. D. (1980). *Nucl. Acids Res.* **8**, 6163–6174.
Shine, J., Fettes, I., Lan, N. C. Y., Roberts, J. L. and Baxter, J. D. (1980). *Nature, Lond.* **285**, 456–461.
Shine, J., Seeburg, P. H., Martial, J. A., Baxter, J. D. and Goodman, H. M. (1977). *Nature, Lond.* **270**, 494–499.
Silman, R. E., Holland, D., Chard, T., Lowry, P. J., Hope, J., Robinson, J. S. and Thorburn, G. D. (1978). *Nature, Lond.* **276**, 526–528.
Smith, A. J. H. (1980). *In* "Methods of Enzymology" (L. Grossman and K. Moldave, eds) vol. 65, 560–580. Academic Press, New York.
Smith, J., Cook. E., Fotheringham, I., Pheby, S., Derbyshire, R., Eaton, M. A. H., Doel, M., Lilley, D. M. J., Pardon, J. F., Patel, T., Lewis, H. and Bell, L. D. (1982). *Nucl. Acids Res.* **10**, 4467–4482.
Stein, S. (1978). *In* "Peptides in Neurobiology" (H. Griner, ed.) 9–37. Plenum Press, New York.
Steiner, D. F., Kemmler, W., Tager, H. S. and Peterson, J. D. (1974). *Fedn. Proc.* **33**, 2105–2115.
Steiner, D. F., Quinn, P. S., Chan, S. J., Marsh, J. and Tager, H. S. (1980). *Ann. N.Y. Acad. Sci.* **343**, 1–16.
Stern, A. S., Jones, B. N., Shively, J. E., Stein, S. and Udenfriend, S. (1981). *Proc. Natn. Acad. Sci. U.S.A.* **78**, 1962–1966.
Sutcliffe, J. G. (1978a). *Nucl. Acids Res.* **5**, 2721–2728.
Sutcliffe, J. G. (1978b). *Cold Spring Harb. Symp. Quant. Biol.* **43**, 77–90.
Svoboda, M. E., Van Wyk, J. J., Klapper, D. G., Fellows, R. E., Grissom, F. E. and Schlueter, R. J. (1980). *Biochemistry* **19**, 790–797.
Taii, S., Nakanishi, S. and Numa, S. (1979). *Eur. J. Biochem.* **93**, 205–212.
Talmadge, K., Brosius, J. and Gilbert, W. (1981). *Nature, Lond.* **294**, 176–178.
Talmadge, K. and Gilbert, W. (1980). *Gene* **12**, 235–241.
Talmadge, K., Kaufman, J. and Gilbert, W. (1980a). *Proc. Natn. Acad. Sci. U.S.A.* **77**, 3988–3992.
Talmadge, K., Stahl, S. and Gilbert, W. (1980b). *Proc. Natn. Acad. Sci. U.S.A.* **77**, 3369–3373.
Tanaka, S., Oshima, T., Ohsue, K., Ono, T., Oikawa, S., Takano, I., Noguchi, T., Kangawa, K., Minamino, N. and Matsuo, H. (1982). *Nucl. Acids Res.* **10**, 1741–1754.
Tatemoto, K. and Mutt, V. (1981). *Proc. Natn. Acad. Sci. U.S.A.* **78**, 6603–6607.
Thayer, R. E. (1979). *Anal. Biochem.* **98**, 60–63.
Thompson, R. (1982). *In* "Genetic Engineering" (R. Williamson, ed.) vol. 3, 1–52. Academic Press, London and New York.
Tu, C.-P.D. and Cohen, S. N. (1980). *Gene* **10**, 177–183.
Twigg, A. J. and Sherratt, D. (1980). *Nature, Lond.* **283**, 216–218.

Ullrich, A., Shine, J., Chirgwin, J., Pictet, R., Tischer, E., Rutter, W. J. and Goodman, H. M. (1977). *Science* **196**, 1313–1319.
Vamvakopoulos, N. C. and Kourides, I. A. (1979). *Proc. Natn. Acad. Sci. U.S.A.* **76**, 3809–3813.
Vamvakopoulos, N. C., Monahan, J. J. and Kourides, I. A. (1980). *Proc. Natn. Acad. Sci. U.S.A.* **77**, 3149–3153.
Villa-Komaroff, L., Efstratiadis, A., Broome, S., Lomedico, P., Tizard, R., Naber, S. P., Chick, W. L. and Gilbert, W. (1978). *Proc. Natn. Acad. Sci. U.S.A.* **75**, 3727–3731.
Wallace, R. B., Johnson, M. J., Hirose, T., Miyake, T., Kawashima, E. H. and Itakura, K. (1981). *Nucl. Acids Res.* **9**, 879–894.
Walter, P. and Blobel, G. (1981). *J. Cell. Biol.* **91**, 557–561.
Wasylyk, B., Derbyshire, R., Guy, A., Molko, D., Roget, A., Teoule, R. and Chambon, P. (1980). *Proc. Natn. Acad. Sci. U.S.A.* **77**, 7024–7028.
Weaver, C. A., Gordon, D. F. and Kemper, B. (1981). *Proc. Natn. Acad. Sci. U.S.A.* **78**, 4073–4077.
Weissmann, C. (1981). *In* "Interferon" (I. Gresser, ed.) vol. 3, 101–134. Academic Press, London and New York.
Wetzel, R., Heynecker, H. L., Goeddel, D. V., Jhurani, P., Shapiro, J., Crea, R., Low, T. L. K., McClure, J. E., Thurman, G. B. and Goldstein, A. L. (1980). *Biochemistry* **19**, 6096–6104.
Wetzel, R., Kleid, D. G., Crea, R., Heynecker, H. L., Yansura, D. G., Hirose, T., Kraszewski, A., Riggs, A. D., Itakura, K. and Goeddel, D. V. (1981). *Gene* **16**, 63–71.
White, B. A. and Bancroft, F. C. (1982). *J. Biol. Chem.* **257**, 8569–8572.
Williams, J. G. (1981). *In* "Genetic Engineering" (R. Williamson, ed.) vol. 1, 1–59. Academic Press, London and New York.
Williams, J. G. K., Shibata, T. and Radding, C. M. (1981). *J. Biol. Chem.* **256**, 7573–7582.
Zimmerman, S. B. and Sandeen, G. (1966). *Anal. Biochem.* **14**, 269–277.
Zinn, K., Mellon, P., Ptashne, M. and Maniatis, T. (1982). *Proc. Natn. Acad. Sci. U.S.A.* **79**, 4897–4901.

Expression of eukaryotic genes in *E. coli*

T. J. R. HARRIS

Celltech Ltd, 250 Bath Road, Slough SL1 4DY, Berks, UK

GENETIC ENGINEERING 4
ISBN 0-12-270304-9

I Introduction

In recent years the techniques of *in vitro* DNA recombination followed by transfection of suitable host cells with recombinant vectors (gene cloning) has led to a great increase in our understanding of the structure and function of the genomes of many organisms. In the early stages of this work it became clear that genes which were cloned in this way could be expressed in the new host if the genetic elements controlling expression were suitably arranged. The results of these efforts will find application in two spheres. In the first, new approaches to fundamental studies on the relationship of protein structure to function will be possible. Already, molecules have been produced which are hybrids of the appropriate regions of different interferon molecules and their functions are being examined. This is possible not only because the genes for the proteins can be recombined but because they can then be expressed in *E. coli* in quantities sufficient for purification and biological study (Streuli *et al.*, 1981; Weck *et al.*, 1981). Further extensions of this kind of work can be foreseen where one or a few selected amino acids (e.g. near the active site of an enzyme) are altered by *in vitro* mutagenesis (Shortle *et al.*, 1981; Lathe *et al.*, 1983) and the effect on enzymatic function assayed. Secondly, such is the power of these gene cloning and expression techniques that they are already having a profound impact on the practice of biotechnology and it seems that few areas of this technology will remain unaffected by them. Indeed, the first proteins made by recombinant DNA techniques are now being produced in sufficient quantity for extensive safety and efficacy testing.

Insulin and growth hormone, both conventionally isolated from human endocrine tissue have now been made in *E. coli* and the proteins purified (Goeddel *et al.*, 1979a, 1979b). Considerable effort has been expended on the isolation and expression of both leukocyte (Le or α) and fibroblast (F or β) interferon genes so that the potential of these antiviral compounds can be evaluated properly (see Scott and Tyrrell, 1980). There is also the possibility of producing proteins for use as vaccines against a variety of infectious agents by cloning and expressing the genes coding for the relevant surface immunogens. Notable progress has been made towards a vaccine for foot and mouth disease virus (FMDV) using this approach, where one of the capsid proteins (VPI) produced in *E. coli* has been shown to elicit neutralizing antibody (Kleid *et al.*, 1981). Genetically engineered vaccines for other viruses such as hepatitis B and rabies virus are also being considered.

Although none of these initial examples of the expression of proteins from recombinant organisms is as yet established as a biotechnological process, the way in which the expression of the recombinant DNA was achieved forms a general paradigm for all future studies. However, at the same time, it is clear that not all the rules governing the expression of cloned genes have been elaborated and those rules that do exist are still largely empirical. In this article the ways in which expression has been achieved are reviewed, some of the problems discussed and some of the probable future systems considered.

II Gene expression in *E. coli*

E. coli has been used as the host cell for expression of foreign genes mainly because more is known about the control of gene expression in this organism than in any other. It is well established, for example, that the genes involved in a particular metabolic activity tend to be clustered in transcriptional units (operons) with the major control regions (the operator and promoter) located at the beginning of the cluster (for a detailed description of bacterial gene expression, see Miller and Reznikoff, 1980). The operon is transcribed into a polycistronic mRNA from which the polypeptides are then translated. Transcriptional control is exerted over the expression of an operon and varies depending on the function of the genes in the operon (see Miller and Reznikoff, 1980). Since relatively few promoter systems are currently being utilized to express cloned genes, the essential elements of their control mechanisms will be dealt with when considering each system. Expression of a cloned gene requires efficient and specific transcription of the DNA, translation of the mRNA and in some cases post-translational modification of the resulting protein.

A Transcription

The first step in the initiation of transcription in *E. coli* is the binding of RNA polymerase to a promoter sequence in the DNA. Analysis of the DNA sequence of many promoters in *E. coli* has revealed two regions of homology located about 35 base pairs (bp) upstream from the transcription initiation site (the −35 region) and about 10 bp upstream (the −10 region or Pribnow-Schaller box). The conserved sequences in the − 35 and − 10 regions (TTGACA and TATAAT respectively, Rosenberg and Court, 1979; Siebenlist *et al.*, 1980) probably represent those bases most intimately involved in polymerase binding and orientation via sigma factor, so that RNA chain initiation can take place just downstream.

Transcription termination is also controlled by signals in the DNA sequence, characteristically a GC rich region having a two-fold symmetry before the termination site, followed by an AT rich sequence at the site of termination (Rosenberg and Court, 1979). Several protein factors are also involved in the control of termination, most notably the rho factor. Anti-termination proteins such as the N gene product of phage λ can also be involved in specialised systems (Greenblatt *et al.*, 1981).

B Translation

Efficient translation of mRNA in prokaryotic cells requires the presence of a ribosome binding site (rbs). For most *E. coli* mRNAs the rbs consists of two components, the initiation codon AUG and, lying 3–12 bases upstream, a sequence of 3–9 bases called the Shine-Dalgarno (SD) sequence complementary to the 3′ end of the 16S rRNA (Shine and Dalgarno, 1975). It is believed that hybridization to this region is involved in the attachment of the ribosomal 30S subunit to the mRNA (Steitz, 1979). The SD sequence is not identical in all mRNAs but a semi-conserved consensus sequence has been identified just as for promoter sequences. It is possible that differences in SD sequences form part of a translational control system. In addition, ribosome binding is probably modulated by the secondary structure at the 5′ end of the RNA since more efficient translation occurs if the AUG and SD sequence are freely accessible to 30S ribosomal subunits (Iserentant and Fiers, 1980). Termination of translation usually occurs whenever one of the three stop codons is encountered in the mRNA by a ribosome complex, provided that an aminoacylated suppressor tRNA is not present.

C Post-translational modification

There are a variety of modifications that bacterial proteins can undergo following translation. The formyl group on the NH_2-terminal methionine is hydrolysed and one or more NH_2-terminal residues may be removed. Many secreted proteins are synthesized as large precursors with additional hydrophobic NH_2-terminal signal sequences that are cleaved off by a membrane bound enzyme (for review, see Davis and Tai, 1980). However, glycosylation and phosphorylation, which are common modifications of proteins in eukaryotic cells do not occur to any great extent in *E. coli*.

III Problems encountered in the expression of eukaryotic DNA in *E. coli*

Successful expression of a eukaryotic gene in *E. coli* requires that the cellular machinery is organised so that the level of expression of the cloned gene is as good or better than the resident genes. Probably the most important difference between eukaryotic genes (at least from higher organisms) and prokaryotic genes is the presence of intervening sequences (introns) which interrupt the coding sequences. Normally these sequences are spliced out of the initial RNA transcript, producing cytoplasmic mRNA suitable for translation. There are no introns in prokaryotic genes and consequently no splicing enzymes present, so in general genomic DNA cannot be used as a source of genes for expression in bacterial cells. A second problem is that transcriptional signals in eukaryotes are different from those in prokaryotes (Corden *et al.*, 1980; Breathnach and Chambon, 1981) and are not usually recognised by bacterial RNA polymerase. This difference again emphasizes the fact that eukaryotic genomic DNA is not a suitable gene source for construction of expression vectors. Thirdly, the structure of eukaryotic mRNA is different to bacterial mRNA. Eukaryotic mRNA is polyadenylated at the 3′ end and normally capped at the 5′ end, features which may affect mRNA stability and ribosome binding (Breathnach and Chambon, 1981). Furthermore eukaryotic mRNA does not seem to have an equivalent of the SD sequence present in prokaryotic mRNA (Kozak, 1981).

An additional problem is that of codon usage. The codons used in mRNA coding for highly expressed prokaryotic genes are not random; there is a marked preference for particular codons for some amino acids (Grantham *et al.*, 1981; see Grosjean and Fiers, 1982). This preference appears to correlate with the abundance of different tRNA species (Ikemura, 1981). As codon selection preferences are different for eukaryotic genes it is possible that the levels of certain tRNAs will affect translational efficiency of these genes in a prokaryotic system. Finally, it is known that many eukaryotic proteins are subject to a number of post-translational modifications which may affect either activity or stability. Most of these modifications do not occur in prokaryotes.

A number of strategies have been developed to try to overcome these difficulties (Table 1). Once the amino acid sequence of a

Table 1 General strategies for the expression of cloned genes in *E. coli.*

Control level	Strategy
Gene	Synthesise DNA *in vitro* by chemical methods, with optimised codon assignments or obtain cDNA clone to specific mRNA. Chemical DNA synthesis probably required for tailoring genes into expression vector.
Transcription (Initiation and termination)	Clone gene adjacent to strong *E. coli* promoter which is controllable so that transcription can be induced (derepressed) when required. Use a multicopy plasmid to increase gene dosage. Include termination signal after gene to prevent transcriptional read-through.
Translation (Initiation)	Fuse gene in correct translational reading frame to an *E. coli* gene already in the vector, so that normal rbs is maintained. Possible to use both long and short NH_2-terminal fusions. Alternatively, place new gene with its own AUG adjacent to an rbs. The sequence of the SD sequence and distance from the initiating AUG may modulate translation. Accessibility (secondary structure) around SD-AUG may be important. Codon usage can be overcome by using chemically synthesized genes. Not clear if codon bias actually affects the translation of cloned genes. Include stop codon(s) in chemically synthesised genes.
Protein (Secretion and stability)	Use signal sequences to control secretion? Synthesis of precursor proteins followed by their processing ensures removal of NH_2-terminal initiating methionine. Factors affecting folding of foreign proteins and their degradation in *E. coli* are not well defined. Synthesis of long fusion proteins may result in increased stability.

protein is known it is now a relatively straightforward task to design and synthesize chemically, a DNA sequence that will code for the protein without the problem of intervening sequences and with optimized codon assignments. A gene of 514 bp coding for leukocyte Le(α) interferon, a protein of 166 amino acids is the longest DNA sequence that has been synthesized so far (Edge *et al.*, 1981). Although there is no theoretical limit to the size of gene that can be synthesized, practical problems arise for much larger proteins. If the gene is too big for a chemical synthesis, then double stranded DNA copies of mRNA populations can be generated, cloned into a plasmid vector and the clone containing the sequence coding for the protein of interest selected from the clone bank by hybridization techniques.

Transcription of these genes is controlled by inserting the DNA adjacent to a strong prokaryotic promoter in a cloning vector. Four promoters have been used most widely for this purpose, the *lac* promoter from the *E. coli lac* operon; the *trp* promoter from the *E. coli trp* operon; the strong leftward promoter of phage λ (P_L) and the constitutive and weaker β-lactamase promoter present in the plasmid vector pBR322. The expression vectors themselves are usually derived from high copy number plasmids so that there is increased expression owing to gene dosage (Gelfand *et al.*, 1978; O'Farrell *et al.*, 1978). Termination of transcription can be ensured by placing a termination site after the cloned gene (e.g. Nakamura and Inouye, 1982) although whether this is necessary for the maintenance of high levels of transcription is not yet clear. The consequences of uninterrupted transcription around a small circular plasmid DNA molecule are unknown. It is presumably detrimental since most expression vectors have other genes present (e.g. an antibiotic resistance gene) which are transcribed in the opposite direction from a different promoter and it is known that the transcription of genes in λ phage carrying the *trp* promoter is adversely affected if the *trp* promoter is in an orientation where transcription occurs towards transcripts arising from the P_L promoter (Hopkins *et al.*, 1976).

Translational barriers have been overcome to some extent by two procedures. The foreign gene is either fused (in the correct translational reading frame) to a prokaryotic gene so that the existing rbs is used to initiate translation, or the new gene, with its own initiation codon, is placed adjacent to a naturally occurring *E. coli* rbs (Backman *et al.*, 1980) or a synthetic one (Jay *et al.*, 1981). Since all structural genes, whether eukaryotic or prokaryotic, end with one or more of the three termination codons it is not usually necessary to make special arrangements for translational termination when using a cloned cDNA sequence. However, a termination codon must be included when synthetic DNA is used.

Protein modification and stability are much less easy to control, largely because the structural features governing protein stability in *E. coli* are not well understood. It has been shown that eukaryotic signal sequences are recognised by *E. coli* and that NH_2-terminal fusions of eukaryotic polypeptides to *E. coli* signal sequences results in secretion of the protein to the periplasmic space, with concomitant cleavage of the signal sequence (Talmadge *et al.*, 1980; 1981). There is also some evidence that short "foreign" polypeptides are unstable in *E. coli* (Itakura *et al.*, 1977; Goeddel *et al.*, 1979a). This has been overcome by fusing the peptide to a larger *E. coli* protein from which the peptide is then cleaved.

IV Expression of DNA from lower eukaryotes

Following the observation that DNA from *S. aureus* could be expressed in *E. coli* (Chang and Cohen, 1974) it was shown that eukaryotic DNA could also be transcribed (Morrow *et al.*, 1974; Chang *et al.*, 1975; Kedes *et al.*, 1975). It was not clear from these experiments, however, whether the normal transcriptional start and stop signals were being recognised. The fundamental question of whether a fungal gene could be transcribed and translated to produce a functional protein in *E. coli* was answered to some extent by the finding that fragments of yeast DNA cloned into phage λ, or the plasmid vector Col E1 could complement auxotrophic mutants of *E. coli* (e.g. His *B* and Leu *B*) (Struhl *et al.*, 1976; Ratzkin and Carbon, 1977; Struhl and Davis, 1977). Similarly segments of *Neurospora crassa* DNA containing the gene for dehydroquinase were successfully expressed in *E. coli* in a pBR322 replicon (Vapnek *et al.*, 1977). Several other yeast genes have now been expressed in this way (e.g. Trp 1, Trp 5 and Arg 4). The functional expression of yeast DNA in *E. coli* not only demonstrated that eukaryotic DNA could be transcribed and translated, paving the way for the experiments described below, but also provided a powerful method for isolating yeast genes. Some of these genes have subsequently been used to provide selection markers in yeast—*E. coli* shuttle vectors (Beggs, 1982; Hinnen and Meyhack, 1982).

The *lac* promoter

The *lac* operon is subject to two types of control. In the absence of lactose (or other inducer) the operon is kept switched off by *lac* repressor (the *lac* i gene product) binding to the operator. Positive regulation is also exerted through the catabolite gene activator protein (CAP). In the absence of glucose, CAP forms a complex with cyclic AMP and this complex stimulates transcription by binding next to the promoter. The operon is derepressed by the presence of lactose, or by the addition of the non-metabolizable inducer IPTG (isopropylthiogalactoside) which binds to the repressor and removes it from the operator.

Plasmid vectors containing parts of the *lac* operon have been constructed by several workers. Polisky *et al.* (1976) cloned an *Eco*RI fragment from λ p *lac* 5 DNA (a transducing phage containing part of the *lac* operon) into a Col E1-derived plasmid to obtain a vector with the *lac* promoter and operator and most of the β-galactosidase gene. Plasmids containing a small "portable" *lac* promoter fragment have also been made. In these constructions a 203 bp *Hae*III fragment of *lac* transducing phage DNA, containing the *lac* promoter and operator and first eight codons of β-galactosidase, was blunt end ligated into *Eco*RI-cut and "filled in" pBR322 DNA. The portability derives from the fact that *Eco*RI sites are reformed at the junctions allowing the promoter fragment to be removed by *Eco*RI digestion (Backman and Ptashne, 1976; Itakura *et al.*, 1977). Colonies harbouring plasmids which carried the *lac* promoter-operator were identified by their constitutive synthesis of β-galactosidase, rendering them blue on agar plates containing X gal (5 chloro-4 bromo 3 indolyl-D galactoside). This is because multiple copies of the operator titrate out all the *lac* repressor resulting in derepression of the chromosomal β-galactosidase gene. Both λ p *lac* 5 DNA and λ h80 *lac* UV5 C_I857 DNA, which contains the CAP site mutation L8 and the up promoter mutation UV5 (making the promoter insensitive to catabolite repression), have been used as a source of *lac* DNA for these constructions (Backman *et al.*, 1976; Itakura *et al.*, 1977; see also Fuller, 1982). Further derivative plasmids containing a 95 bp *Alu*I fragment of *lac* DNA, including the UV5 promoter (minus the CAP binding site), the repressor binding site and most of the rbs, just excluding the ATG of β-galactosidase, have also been constructed for the expression of non-fusion proteins (Fuller, 1982).

A The somatostatin experiment

The first report of the designed expression of a eukaryotic gene in *E. coli* was the production of the small peptide hormone somatostatin (Itakura *et al.*, 1977). Somatostatin was used as a model system because the hormone was a small polypeptide of known amino acid sequence for which sensitive radioimmune and biological assays existed. The experiments illustrate a number of features of methods which are now used to obtain expression of cloned genes. They also demonstrated, although not for the first time, that chemically synthesized DNA was functional in a biological system. In addition, the production of the protein as a fusion polypeptide and its subsequent cleavage into the native hormone at methionine residues by cyanogen bromide (CNBr), has been used quite extensively for other proteins. This overall strategy is depicted in Fig. 1.

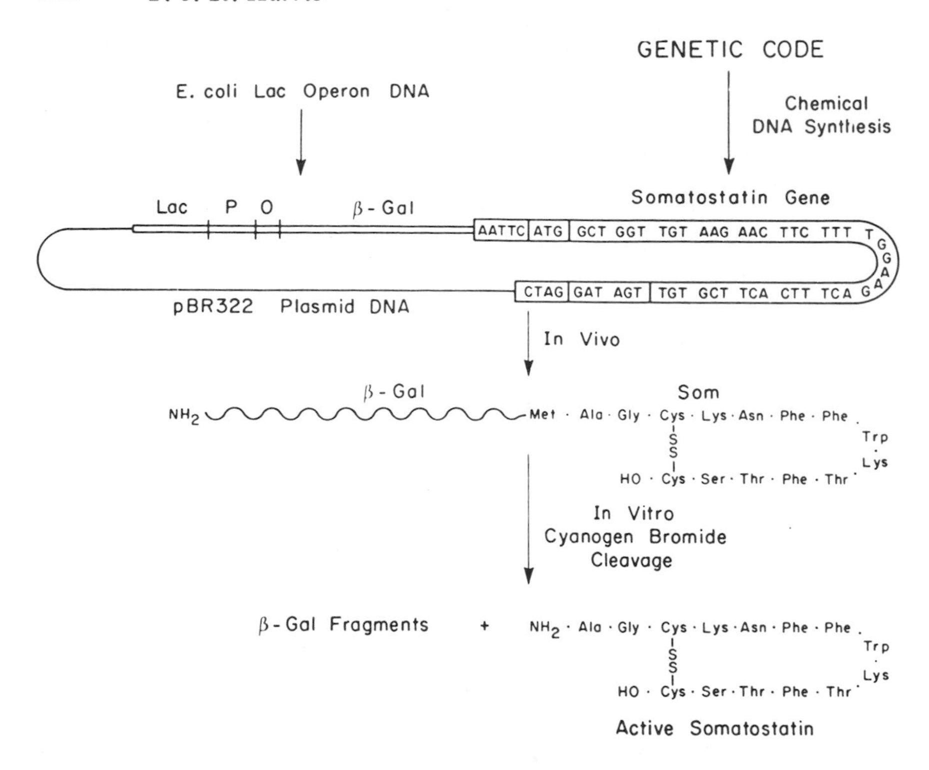

Figure 1 Strategy for the expression of the chemically synthesized somatostatin gene as a β-galactosidase fusion from the *lac* promoter. The active hormone can be cleaved from the hybrid protein by CNBr treatment. (Reproduced from Itakura *et al.*, 1977, copyright by the American Association for the Advancement of Science, with permission.)

In the first set of experiments the chemically synthesized somatostatin gene with synthetic *Eco*RI and *Bam*HI cohesive ends was cloned into a vector containing the wild type *Hae*III *lac* promoter fragment. The DNA sequence indicated that the plasmid should have produced a polypeptide containing the first seven amino acids of β-galactosidase fused to somatostatin. However, no somatostatin was detected in bacterial extracts by radioimmunoassay. As it was found that somatostatin was not stable when added to *E. coli* extracts, the failure to find somatostatin was thought to be due to proteolytic digestion (Itakura *et al.*, 1977). The approach adopted to try to stabilise the somatostatin was to produce it as part of a longer polypeptide from which it could be cleaved by CNBr. This was done by linking the somatostatin gene to the *Eco*RI fragment of λ p *lac* 5 DNA which carries the *lac* promoter and a large proportion of the β-galactosidase gene (Polisky *et al.*, 1976). The translation reading frame of β-galactosidase was maintained in somatostatin after fusion at the *Eco*RI junction. In these constructions only one orientation of the *Eco*RI *lac* fragment maintained the correct reading frame in somatostatin and when several independent clones were examined, about half produced detectable somatostatin after CNBr cleavage. No immunoreactive protein was detected before cleavage since the antiserum used in the assay required a free NH_2-terminal alanine residue (Itakura *et al.*, 1977).

B Expression of insulin in *E. coli*

The somatostatin work established the feasibility of the synthetic gene fusion approach for the expression of small polypeptides in *E. coli.* It was possible to follow an almost identical strategy to obtain expression of human insulin, as neither the 20 amino acid A chain nor the 30 amino acid B chain of insulin contained methionine and methods were available for the *in vitro* joining of the two chains. Thus, an A chain gene and a B chain gene were chemically synthesized each with *Bam*HI and *Eco*RI cohesive ends (Crea *et al.*, 1978) and cloned separately into pBR322. The B chain gene was synthesized with a *Hind*III site in the middle so that the two halves could be cloned separately and the sequence verified (Goeddel *et al.*, 1979a). Expression was achieved by transcription from the same *lac* promoter as used for the successful somatostatin constructions and insulin A or B-β-galactosidase fusion proteins were produced (Goeddel *et al.*, 1979a). The hybrid proteins represented about 20% of total cell protein, which was about ten-fold higher than the level of expression obtained with somatostatin. The hybrid proteins were insoluble and were found in the first low speed pellet after breaking the cells with a French press.

To obtain A and B peptides suitable for reconstitution into native insulin, the hybrid proteins had to be solubilised, the β-galactosidase portion removed and the peptides S-sulphonated. This was achieved by dissolving the hybrid proteins in 6 M guanidinium chloride followed by dialysis. The precipitate was dissolved in 70% formic acid, the protein cleaved with CNBr and S-sulphonated derivatives of the peptide mixture obtained, using sodium dithionate and sodium sulphite at pH 9. Insulin activity was readily detected by radioimmunoassay after re-constitution. Further studies on the peptides (e.g. chromatographic behaviour) and amino acid compositions established, without doubt, that the bacteria were producing authentic insulin A and B chains (Goeddel *et al.*, 1979a). Insulin, prepared from bacteria containing these constructions by a scaled up and modified process, has now been shown to be active when injected into human volunteers (Clark *et al.*, 1982) and to interact with insulin receptors in the same way as native human insulin (Keefer *et al.*, 1981).

An alternative approach involves the synthesis of a gene coding for proinsulin, the natural precursor to insulin. Proinsulin is synthesized initially as a preproinsulin molecule consisting of an NH_2-terminal signal sequence, followed by the B chain, a linking C peptide and the COOH-terminal A chain. Enzymatic removal of the signal peptide during secretion generates proinsulin and processing at two trypsin sensitive sites (Arg-Arg, Lys-Arg) allows the removal of the C peptide and the generation of active insulin. The three dimensional structure of insulin indicates that a peptide much shorter than the 35 amino acid connecting C peptide should be sufficient to connect the B and A chains and still allow proper folding of the modified proinsulin. Genes coding for human proinsulin and "mini C" derivatives of proinsulin, where the C peptide is replaced by a six amino acid linker retaining the proteolytic cleavage sites, have been constructed by chemical synthesis (Sung *et al.*, 1979; Wetzel *et al.*, 1981a; Brousseau *et al.*, 1982).

The mini C construction was cloned for expression as a β-galactosidase fusion protein (Wetzel *et al.*, 1981a) and a product with a proinsulin-like structure (as determined by radioimmunoassay and HPLC) was detected after CNBr cleavage and S-sulphonation. The usefulness of this route to insulin production is still not clear however, as there are no data on the behaviour of mini C derivatives in enzymatic proinsulin processing systems and there are already preproinsulin expression constructions available derived from cDNA (see β-lactamase section). However, the modular approach to the chemical synthesis of proinsulin adopted by Brousseau *et al.* (1982) does have the advantage that the shortening and changing of parts of the C peptide or alteration of the codons can be approached rationally by the incorporation of different oligonucleotide blocks during synthesis, obviating the need to synthesise an entire coding sequence each time a specific modification is made.

C Synthesis of other hormones as β-galactosidase fusions

The strategy of using the *lac* promoter/operator and β-galactosidase NH_2-terminal fusions has been adopted for several other proteins including other hormones (see Table 2). For example the neuropeptide β-endorphin, a 30 amino acid endogenous opioid has been expressed in this way (Shine *et al.*, 1980). In these experiments a cDNA clone to the precursor peptide of mouse corticotropin (ACTH) and β-lipotropin (LPH) was used as a source of cDNA coding for

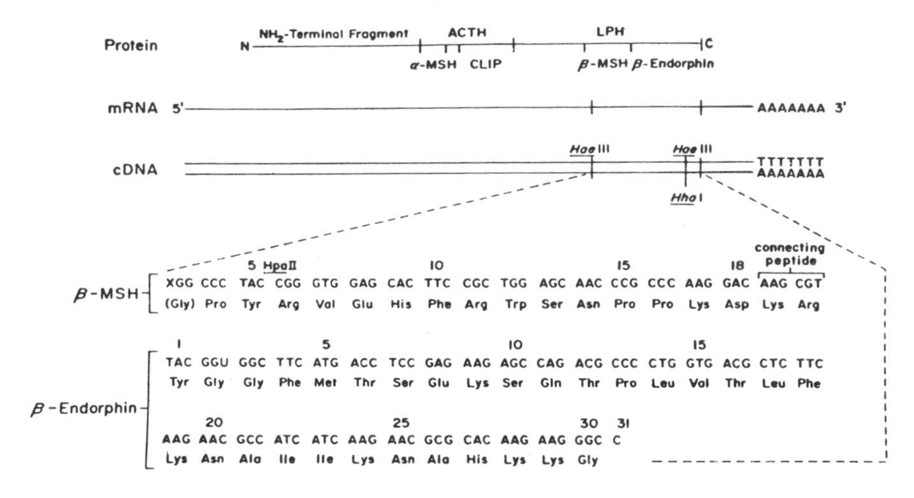

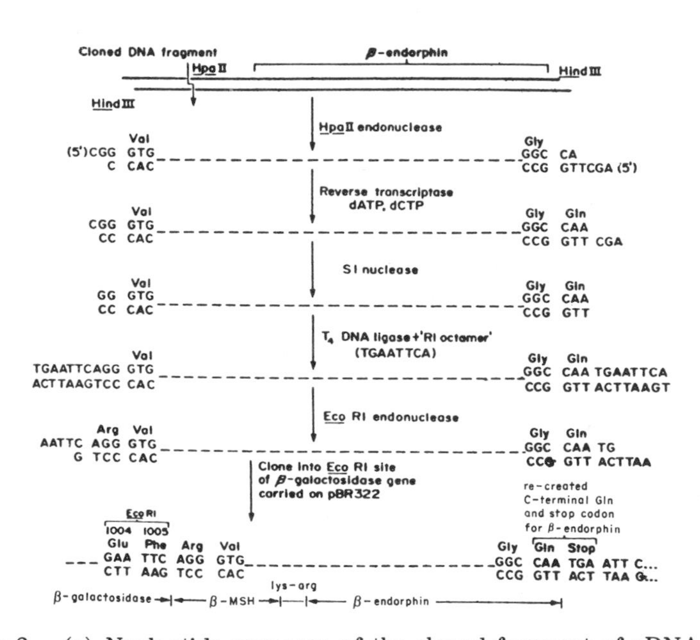

Figure 2 (a) Nucleotide sequence of the cloned fragment of cDNA to ACTH β-endorphin mRNA (b) β-galactosidase fusion construction used for the expression of the hybrid β-endorphin protein. Reprinted by permission from Shine *et al.*, *Nature* **285**, 456--461. Copyright © 1980, Macmillan Journals Ltd.

β-endorphin (Roberts *et al.*, 1979a). The cDNA fragment contained all the coding sequence except the C-terminal glutamine. For the expression constructions it was necessary to recreate the C-terminal codon, insert a stop codon and link the gene in phase to β-galactosidase. This process is illustrated in Fig. 2. The cloned

Hind III fragment containing the β-endorphin coding sequence was cleaved with *Hpa* II at one end and the cohesive ends filled in using reverse transcriptase in the presence of dATP and dCTP, so that only partial filling in occurred. This step regenerated the C-terminal glutamine. The remaining overhanging ends were then removed with S1 nuclease and the fragment blunt end ligated to a chemically synthesised linker containing a stop codon and an *Eco*RI site (Fig. 2). Subsequent *Eco*RI digestion generated a fragment which could be linked in phase to β-galactosidase. As previously found for the insulin constructions, the hybrid β-endorphin/β-galactosidase hybrid protein was insoluble but represented a substantial proportion of total protein in the pellet obtained after disruption and centrifugation (Shine *et al.*, 1980).

Since β-endorphin contains a methionine residue at amino acid 5, CNBr cleavage could not be used to cleave the hybrid protein. An alternative strategy was developed based on the fact that an arginine residue, which is a site for trypsin cleavage, is present in β-melanocyte stimulating hormone, the peptide preceding β-endorphin, but not in β-endorphin itself and that the lysine residues in β-endorphin can be protected from trypsin attack by citraconylation (which is reversible). Thus, after dissolving the precipitated hybrid protein and treatment with citraconic anhydride at pH 9, the modified β-endorphin was cleaved from the hybrid protein by trypsin digestion. Various immunological and biological criteria showed that authentic active murine β-endorphin had been synthesized (Shine *et al.*, 1980). Since the murine protein differs from the human protein in only two positions (tyr for his at position 27, glu for gln at position 31) it is possible that a clone expressing human β-endorphin could be made by altering the codons in these two places, using oligonucleotide site directed mutagenesis (Smith and Gillam, 1981). It seems unlikely however, that a trypsin cleavage protocol of this kind could be economically or efficiently used on a large scale.

D Expression of ovalbumin

The *lac* promoter has also been used to express an oval bumin cDNA clone. Charnay *et al.* (1979) cloned the promoter fragment back into λ p *lac* 5-1 giving a molecule with two *lac* regulatory regions. In one orientation intramolecular recombination occurred generating a phage with only one *Eco*RI site downstream from the *lac* promoter, after the 8th codon of β-galactosidase (λ p *lac* UV5, Charnay *et al.*, 1979). The *lac* control region from this phage was transferred back to pBR322 to create a plasmid (pOMPO) with an *Eco*RI site into which was blunt end ligated a *Hha* I fragment of an ovalbumin cDNA clone (Mercereau-Puijalon *et al.*, 1978). In the correct orientation, the translation frame was maintained to generate an ovalbumin fusion protein where the NH_2-terminal five amino acids of ovalbumin were replaced by the NH_2-terminal eight amino acids of β-galactosidase. An ovalbumin-like protein was detected by radioimmunoassay and by polyacrylamide gel electrophoresis of immunoprecipitates of ^{35}S methionine labelled bacterial extracts. In a bacterial strain which overproduces *lac* repressor (repressing the multiple copies of the operator) the synthesis of the hybrid protein was stimulated over 50-fold by the addition of IPTG, showing that the synthesis was under *lac* control. Similar results were reported by Fraser and Bruce (1978), although in their construction the fusion protein contained an additional 18 amino acids at the NH_2-terminus, and was apparently secreted into the periplasmic space. Baty *et al.* (1981) have confirmed that ovalbumin made as a short β-galactosidase fusion protein in *E. coli* is transported to the periplasmic space and have shown further that a derivative of the protein lacking the NH_2-terminal 126 amino acids remains in the cytoplasm.

These were the first reports of the synthesis of a large (> 40 000 molecular weight) eukaryotic protein in *E. coli*. Although there did not appear to be any overriding stability problems for either the plasmid containing the gene, or for the protein, only about 10% of the theoretical yield of ovalbumin (based on native β-galactosidase synthesis) was produced. This could have been due to inefficient transcription or translation or to proteolytic degradation. Another possible reason for the low level of expression was that the codons in the gene derived from ovalbumin mRNA were not optimal for *E. coli* and the availability of certain minor isoaccepting tRNAs could therefore have been limiting (Mercereau-Puijalon *et al.*, 1978).

The production of ovalbumin in *E. coli* was obviously dependent on maintaining the correct translation frame across the β-galactosidase-ovalbumin junction. Charnay *et al.* (1978) constructed a set of plasmid and λ phage vectors allowing fusion of cloned genes in each of the three translational phases. This was achieved by treating the *Eco*RI *lac* UV5 promoter fragment with S1 nuclease and ligating a synthetic octanucleotide linker, containing an *Eco*RI site, to the blunt ends. Cutting with *Eco*RI generated a fragment with two additional base pairs before the cohesive end. Repeating the procedure generated a fragment with a further two base pairs. Considering the original vector to give translation in frame 1, then the two new constructs will give translation frame 2 and frame 3 respectively when fused to the same *Eco*RI fragment (Charnay *et al.*, 1978; 1979).

E Expression of native proteins

In all the constructions examined so far the proteins have been produced from fusions with parts of β-galactosidase. Backman and Ptashne (1978) extended their studies with the λ repressor by making a construction where the λ C_1 gene was fused to the β-galactosidase rbs, rather than within the coding region, forming a hybrid β-galactosidase-C_1 gene SD sequence 8 bases from the AUG of the λ repressor. This construction led to the synthesis of about 30 000 molecules of λ repressor per cell; other fusions which placed the *lac* promoter further away from the C_1 gene reduced the synthesis of repressor quite considerably. Much larger amounts of protein were produced when *lac* rbs fusions were made to the λ *cro* gene by the same group (Roberts *et al.*, 1979b; Lauer *et al.*, 1981). This has allowed a more systematic analysis to be carried out of the effect on expression of changing the distance between the promoter and the gene (Roberts *et al.*, 1979b).

A series of plasmids were constructed in which the λ *cro* gene was placed at varying distances downstream from the *lac* promoter and SD sequence. The level of *cro* protein produced by clones containing different plasmids was then correlated with the nucleotide sequence across the *lac-cro* junction. There was considerable variation in the amount of *cro* produced by the various fusions (Fig. 3). Strains containing plasmids pTR213 or pTR214, produced about 200 000 monomers of *cro* per cell (1.6% total cell protein) calculated to be the number expected from a fully induced *lac* promoter, whereas pTR199 with almost the same number of nucleotides between the *lac* SD and the *cro* AUG directed the synthesis of only about 1/10th as much *cro* protein. Since the same promoter was being used in all the plasmids it was concluded that some post-transcriptional process was responsible for the differences in level of *cro* synthesis. It was unlikely to be caused by plasmid copy number variability nor by the regulatory effects of the *cro* protein itself, since the operators at which *cro* acts were not present (Johnson *et al.*, 1981). An explanation for the observations of Roberts *et al.* (1979b) has been advanced based on secondary structure models for the 5′ terminus of these different *cro* mRNAs (Iserentant and Fiers, 1980). The involvement of the AUG and the SD sequence in secondary structure is thought to reduce the amount of protein produced, presumably because interaction with the 30S ribosomal subunit is inhibited. Sequences which produce mRNAs with secondary structure that leave the AUG accessible (e.g. pTR213) produced large amounts of protein. Involvement of the SD sequence in base paired regions apparently caused far smaller reductions in *cro* protein

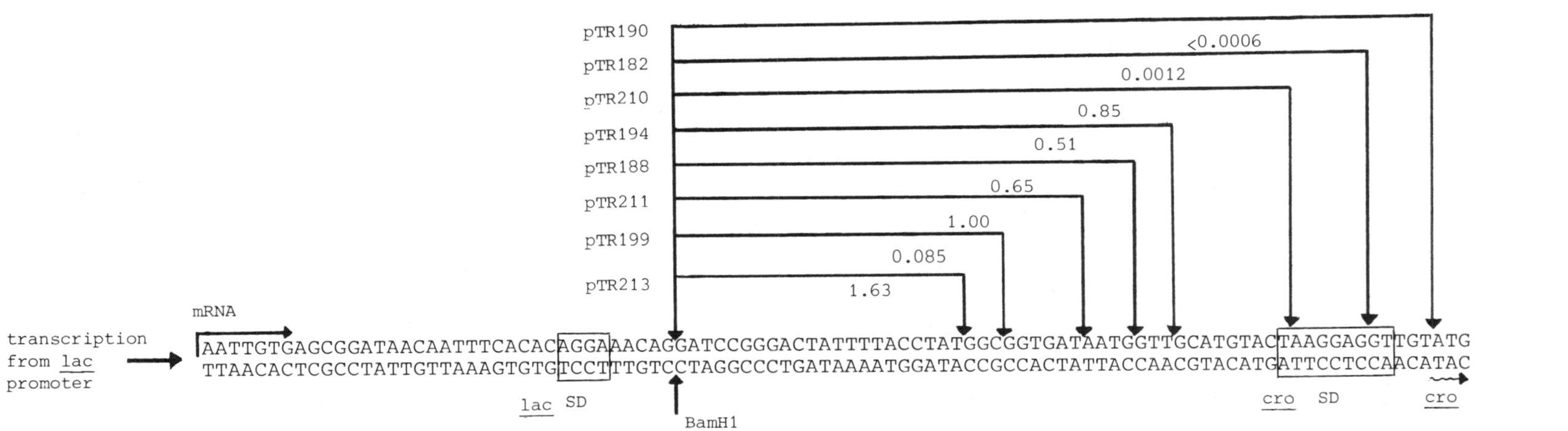

Figure 3 Correlation of nucleotide sequence deletion limits and *cro* protein production for various hybrid rbs constructions between the *lac* SD, the *cro* SD and the *cro* initiation codon. Adapted from Roberts *et al.*, 1979b, by permission.

synthesis than loss of an accessible AUG (Iserentant and Fiers, 1980).

Secondary structure in mRNA is dictated not only by the distance between the SD and AUG but also by the sequence between them and the sequence coding for the remaining N-terminal amino acids. It is not therefore, a trivial task to predict the sequence and optimal distance between the promoter and the AUG to obtain maximal expression of a particular gene. With this fact in mind, Guarente *et al.* (1980) have devised a general method for maximizing the expression of cloned genes in the absence of assays for the gene products. In this method a fragment of DNA bearing the NH_2-terminal region of a gene is fused to a DNA fragment coding for the enzymatically active COOH-terminal fragment of β-galactosidase in an analogous manner to that used to examine the control of prokaryotic promoters (Casadaban *et al.*, 1980; see also Bassford *et al.*, 1978). The portable *lac* promoter is then placed at varying distances in front of the fusion. The constructions which lead to efficient expression of the fused gene are recognised by the amount of β-galactosidase expressed. The gene is then reconstituted in the plasmids directing high levels of β-galactosidase, where an optimal promoter-SD-ATG distance should have been established (Fig. 4). The method was tested with β-globin cDNA and has since been used to obtain expression of a human fibroblast interferon (HFIF) cDNA gene (Guarente *et al.*, 1980; Taniguchi *et al.*, 1980). Plasmids were obtained directing the synthesis of HFIF and the precursor protein pre-HFIF. It was interesting that in each of these constructions the distance separating the SD sequence of the *lac* promoter and the ATG was found to be precisely that normally found between the ATG of β-galactosidase and the *lac* SD sequence. The levels of protein produced were not high, possibly owing to proteolytic degradation (Taniguchi *et al.*, 1980). Other vectors based on the *lac* promoter allowing fusion of DNA coding for enzymatically active β-galactosidase to that coding for amino-terminal fragments of exogenous proteins have also been constructed (Casadaban *et al.*, 1980).

F Expression of human growth hormone

The approach used for the expression of human growth hormone (HGH) as a native rather than a fused polypeptide demonstrates the combination of chemical synthesis and cDNA cloning (Goeddel *et al.*, 1979b). The specific strategy was based on the known restriction map of HGH cDNA. Treatment of cloned HGH cDNA with *Hae*III generated a 551 bp fragment coding for amino acids

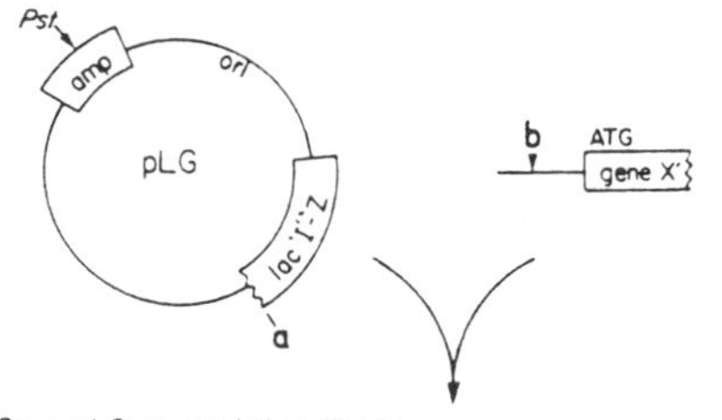

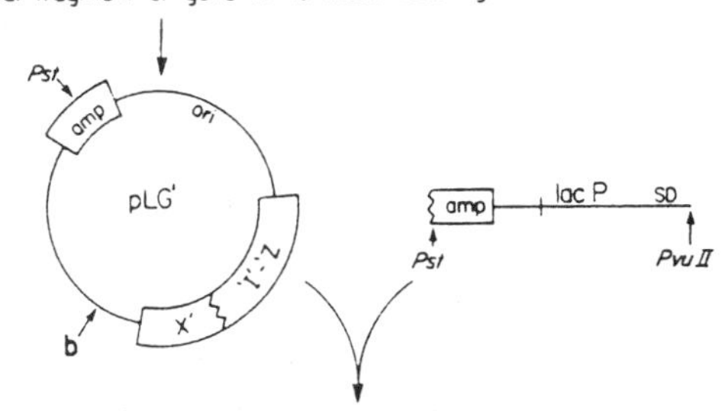

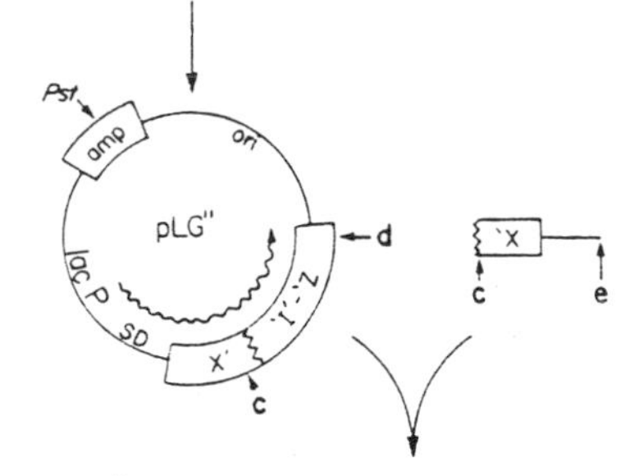

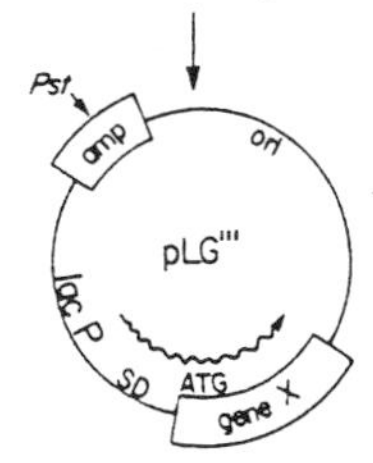

Figure 4 A general method for maximizing expression of a cloned gene in *E. coli.* Reproduced from Guarente *et al.*, 1980, *Cell* **20**, 543–553, by permission from MIT press.

24–191 of HGH (Fig. 5). An adaptor fragment encoding an initiation codon and amino acids 1–24 (including the *Hae* III site found in the cDNA) was chemically synthesized. The two fragments were cloned separately, the cDNA by C tailing into the *Pst* I site of pBR322 and

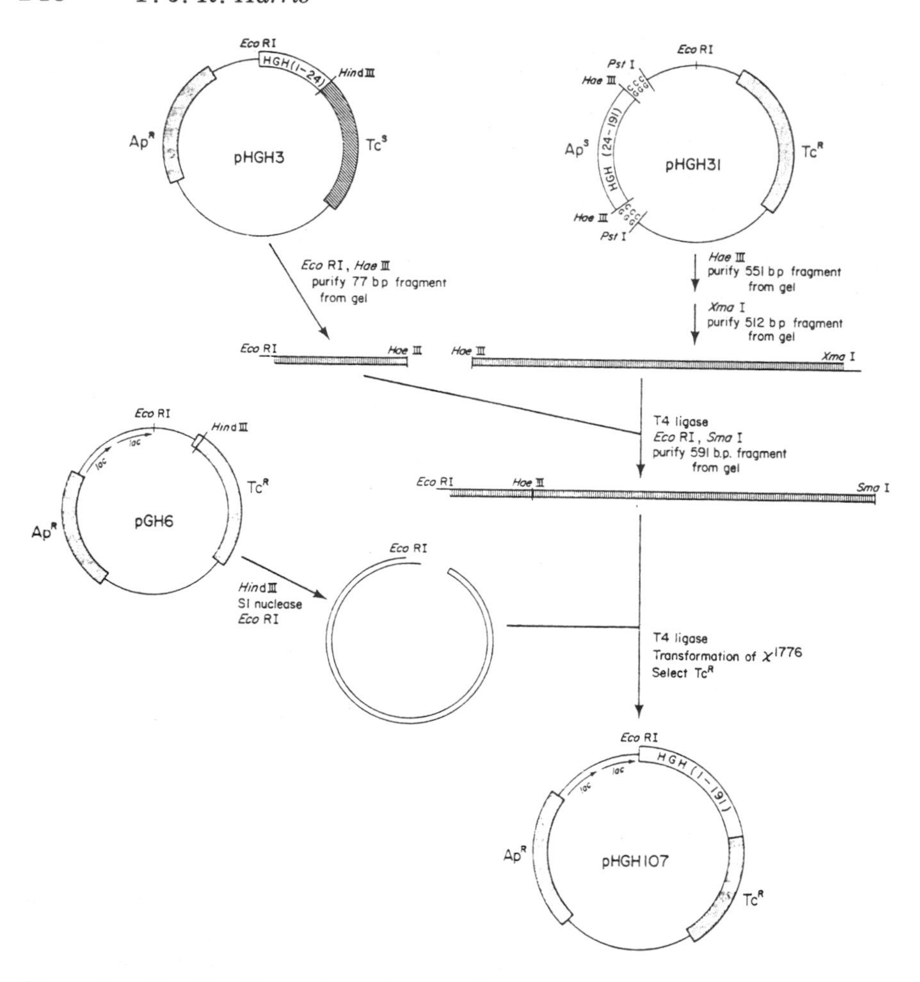

Figure 5 Construction of a plasmid for the expression of human growth hormone in *E. coli* (Goeddel *et al.*, 1979b). Reprinted by permission from *Nature* 281, 544–548. Copyright © 1979 Macmillan Journals.

the adaptor fragment into pBR322 as an *Eco*RI-*Hind*III fragment. Two fragments were isolated from these vectors (Fig. 5), a 77 bp *Hae*III/*Eco*RI fragment from pHGH3 and a 512 bp *Hae*III-*Xma*I fragment from pHGH31. The two fragments were ligated, treated with *Eco*RI and *Sma*I and the 591 bp. DNA coding for HGH isolated and inserted into an expression vector containing two copies of the *lac* promoter (Fig. 5). One of the resulting plasmids (pHGH 107) was found by nucleotide sequence analysis to have 11 base pairs separating the *lac* SD sequence and the ATG for HGH. This was reduced to seven base pairs (the naturally occurring distance for *lac* SD and β-galactosidase) by *Eco*RI digestion, S1 nuclease treatment and blunt end ligation to generate a new plasmid (pHGH107.1) (Goeddel *et al.*,

1979b). Unexpectedly, this new construction produced less immunoreactive HGH than pHGH107 with the longer ATG-SD distance. Growth hormone was readily detectable in polyacrylamide gels of the proteins of extracts of *E. coli* RV 308 (a nutritionally wild type K12 strain) carrying pHGH107.

In contrast to the insulin chain and the β-endorphin fusion polypeptides, HGH is produced as a soluble protein. Partial purification was achieved by ammonium sulphate precipitation and Sephacryl S-200 gel filtration. Preparations of HGH of high purity have now been obtained from *E. coli* carrying these plasmids and this HGH has been shown to have the same specific activity as natural HGH derived from human pituitary glands (Olson *et al.*, 1981; Hintz *et al.*, 1982; Rosenfeld *et al.*, 1982). Very precise authenticity studies have also been carried out on the *E. coli* product; HPLC of the protein and its constituent tryptic peptides has shown that bacterial HGH has the same amino acid sequence and disulphide bond arrangement as the natural hormone apart from the presence of an extra NH_2-terminal methionine (Kohr *et al.*, 1982). The possible side effects of the presence of this methionine (e.g. antigenicity) is a current cause for concern. As this will be a universal problem for bacterially derived products which do not naturally retain an initiating methionine residue, it may be necessary to devise methods for its specific removal.

VI The phage λ P_L promoter

The observation that the *trp* operon genes of *E. coli* could be expressed in phage λ derivatives under the control of the leftward promoter P_L (Moir and Brammar, 1976) suggested that this promoter could be used to drive expression of cloned DNA. In the normal phage infection cycle the P_L promoter controls early leftward transcription of the DNA through gene N to *int* (for review, see Szybalski and Szybalski, 1979). P_L is a strong promoter which is subject to various forms of control. Most importantly the promoter is subject to repression by the C_I gene product and later in infection by the *cro* protein. There is also the *Nut* L sequence downstream from the promoter which allows the N gene product, in association with RNA polymerase, to overcome transcription termination further downstream.

It is possible to obtain expression from P_L either by inserting DNA into the phage itself or by cloning the promoter into a suitable plasmid vector. There have been several reports of the use of hybrid λ phages for expressing prokaryotic genes. In most of these studies the genes have been cloned with their own promoter and oversynthesis occurs primarily due to the increased copy number attained by phage DNA replication (Panasenko *et al.*, 1977; Hopkins *et al.*, 1976; Kelley *et al.*, 1977). Studies with the *E. coli* DNA polymerase gene (pol A) cloned in a λ phage derivative under P_L control have illustrated some of the problems concerning the use of the P_L promoter within a phage DNA molecule. For example the *cro* gene product, which is required for efficient DNA replication, represses transcription from P_L preventing sustained transcription (Murray and Kelley, 1979). Both P_L and other promoters such as the late rightwards promoter (P'_R) have, however, been used to obtain effective expression of tryptophan synthetase (Moir and Brammar, 1976) and T4 DNA ligase (Murray *et al.*, 1979).

Very few studies have been done concerning the expression of eukaryotic genes from the P_L promoter within phage λ. Kourilsky *et al.* (1977) reported that β-globin cDNA is expressed when cloned into the early region of the phage and low levels of interferon have been detected in *E. coli* infected with a Charon 4A recombinant phage containing a fragment of the human genome (Mory *et al.*, 1981). In addition, the wheat chloroplast gene for the large subunit of ribulose bisphosphate carboxylase has been expressed from P_L when cloned into a phage λ derivative (Gatenby *et al.*, 1981).

Most of the studies have been done using cloned λ promoters because the problem of control of expression of the promoters (by using mutations in other phage genes) can be largely avoided if the promoter itself is cloned in a plasmid vector. One of the potential advantages of the P_L promoter is that, in contrast to the *lac* promoter, there is sufficient λ repressor produced from a single copy of the C_I gene to repress transcription from multiple plasmid-borne copies of P_L. By using a temperature sensitive C_I gene (C_I857), it is possible to control transcription from P_L such that at 28°C it is entirely switched off but at 42°C it is fully induced. (The C_I857 gene, maintained in the host chromosome or in a compatible plasmid, produces a repressor which is inactivated at 42°C.) As an example, Bernard *et al.* (1979) showed that heat induction of plasmids containing the *trp* A gene downstream from P_L, controlled by C_I857, produced up to 6.6% of soluble cell protein as tryptophan synthetase.

Remaut *et al.* (1981) have also described plasmid expression vectors based on λ P_L. All these vectors incorporated a 247 bp DNA fragment from the phage containing the operator-promoter region of P_L and 114 nucleotides of the P_L transcript, excluding the initiation site for N protein, but including the *Nut* L site. Unique restriction sites for *Eco*RI, *Bam*HI and *Hind*III were present further downstream (Fig. 6). P_L activity was controlled by maintaining the plasmids in partial λ lysogens containing a chromosomal C_I857 gene but no *cro*

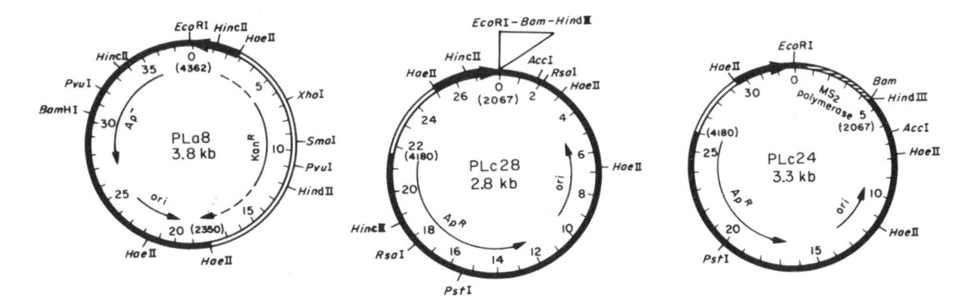

Figure 6 Restriction and genetic maps of representative P_L vectors. The position of the O_LP_L region present on the vectors is indicated by a heavy arrow showing the direction of transcription. The heavy solid lines represent sequences derived from pBR322. Ap^R indicates the region coding for β-lactamase. The direction of translation is shown by an arrow. Kan^R indicates the region carrying the resistance to kanamycin. Ori indicates the region of the origin of replication. The direction of replication is shown by an arrow. Only some particularly relevant restriction sites are shown. The small *Hae*II fragment constituting part of the origin of replication originated from Col E1 in the case of PLa plasmids and from pBR322 in the case of PLc plasmids. Taken from Remaut *et al.* (1981) by permission.

gene (K-12ΔHIΔtrp). In one of the strains (M5219) a functional N gene was also present in the chromosome providing an additional potential control in *trans* by antitermination.

In strains with P_L plasmids containing the β-lactamase gene from pBR322 or the *trp* A gene of *E. coli*, synthesis of high levels of β-lactamase and tryptophan synthetase (up to 10% of total cell protein) could be demonstrated, particularly when the promoter was cloned into the pBR322 replicon and transcribing in the clockwise direction in M5219 (see Fig. 6) (Remaut *et al.*, 1981). This was a somewhat higher level of *trp* A than that reported by Bernard *et al.* (1979) where N protein was expressed from the plasmid rather than from the chromosome.

Large amounts of the λ C_{II} protein, a phage regulatory protein toxic to *E. coli*, have also been produced by vectors containing the P_L promoter (Shimatake and Rosenberg, 1981). The C_{II} gene was cloned into a *Hpa*I site located 321 bp downstream from the P_L transcription start site. In the correct orientation a λ lysogen could be transformed at high efficiency, whereas cells not making λ repressor could not be transformed. In strains containing a chromosomal C_I857 gene, the plasmid directed the synthesis of about 4% total cell protein as the C_{II} gene product on temperature shift induction.

A Expression of eukaryotic genes from P_L plasmids

Vectors based on P_L have been used to express a variety of eukaryotic genes. Derynck *et al.* (1980) cloned two HFIF cDNA genes into the vector containing the β-lactamase gene (pPLa8, Fig. 6) to produce a plasmid (pPLa HFIF-67-12Δ19) containing the HFIF cDNA sequence fused in phase to the β-lactamase gene. The predicted hybrid protein contained 83 amino acids of β-lactamase followed by the methionine of the signal sequence of HFIF. In a second construction (pPLc HFIF-67-8) the acceptor plasmid was pPLc24 (Fig. 6) which contains the P_L promoter followed by an *Eco*RI-*Bam*HI fragment containing the rbs and part of the MS2 phage RNA polymerase gene. In this case, the expected fusion protein consisted of the N-terminal 98 amino acids of MS2 polymerase, 27 amino acids coded for by the linking sequences, followed by the methionine of the signal sequence of HFIF. Antiviral activity could be detected in cleared lysates of bacterial strains containing both constructions after induction at 42°C. Lysis with a mixture of SDS, β-mercaptoethanol and urea resulted in a 10–20 fold increase in activity (HFIF is stable under these conditions), suggesting that some non-specific protein aggregation was occurring. Although the levels of synthesis were low, the solubilized polypeptides had physicochemical, biological and immunological properties resembling HFIF. Polyacrylamide gel electrophoresis suggested, in addition, that some post-translational cleavage of the fusion proteins had occurred generating native HFIF (Derynck *et al.*, 1980).

The small t antigen (tAg) of SV40 has also been expressed in these vectors (Derom *et al.*, 1982). This protein was chosen largely because the role of the tAg in SV40 transformation was equivocal and insufficient protein was available from SV40 transformed cells for detailed biochemical study. In these constructions a DNA segment functioning as an rbs was inserted between P_L and the tAg gene so that a native rather than a fusion protein was produced. The best construction expressed 2.5% of *de novo* protein synthesis as authentic tAg. This level is better than the level of tAg produced in *E. coli* from vectors containing the tAg gene fused to the *lac* promoter, and utilizing a hybrid rbs (Roberts *et al.*, 1979c; Thummel *et al.*, 1981).

In addition to the 19 000 molecular weight small t protein several of the plasmids expressed a related 14 500 molecular weight polypeptide (Derom *et al.*, 1982). Shorter forms of small t were also found in bacteria containing *lac*-small t fusions (Roberts *et al.*, 1979c; Thummel *et al.*, 1981). Tryptic peptide maps indicated that the shortened polypeptides lacked N-terminal peptides suggesting that they arose by initiation at an internal AUG (Thummel *et al.*, 1981;

Derom *et al.*, 1982). In agreement with this conclusion secondary structure maps of the mRNA transcribed from plasmids producing the short tAg indicated the presence of a freely accessible AUG (Gheysen *et al.*, 1982). Systematic alteration of the distance and/or nucleotide sequence between the SD sequence and the AUG of tAg in these plasmids has provided additional evidence that efficient initiation of translation requires an accessible AUG and SD sequence (Gheysen *et al.*, 1982; see also Iserentant and Fiers, 1980).

VII The *trp* promoter

A Construction of vectors

Hershfield *et al.* (1974) showed that high levels of expression of the five genes of the *trp* operon occurred when the operon was transferred to a plasmid vector (for a review of the more molecular aspects of the *trp* operon, see Yanofsky *et al.*, 1981). With these plasmids, the *trp* enzymes accounted for about 20—25% of total cellular protein after induction with 3-indolylacrylic acid (IAA). Moreover, in contrast to the *lac* system, in the absence of inducer, a single copy of the *trp* repressor gene produced sufficient protein to keep the operon fully repressed, despite the increased operator copy number. These observations led to the construction of expression vectors where genes are inserted under the control of the *trp* operon regulatory elements (Hallewell and Emtage, 1980). A 5.4 kb *Hind* III fragment of *E. coli* containing the *trp* promoter, operator, leader and attenuator, all of the *trp* E gene and part of the *trp* D gene was cloned into the *Hind* III site of pBR322 to form p *trp* ED3. The *Hind* III site near the single *Eco*RI site of p *trp* ED3 was subsequently removed using exonuclease III and nuclease S1 and a deleted plasmid isolated (p *trp* ED5-1, Fig. 7) having a *Hind* III site at the end of the *trp* D gene suitable for insertion of cloned genes. Polyacrylamide gel electrophoresis of the proteins made in cells containing p *trp* ED5-1, 3 h after induction with IAA, indicated that about 30% of total cell protein was anthranilate synthetase. This was more protein than was made by similar *lac* based vectors containing the *lac* z gene and furthermore the synthesis was inducible rather than constitutive. In addition, it was shown that the small protein derived from the deleted *trp* D gene (15% of full length) was stable, probably because it was bound to the *trp* E protein (Hallewell and Emtage, 1980).

B Expression of fusion proteins

Plasmid p*trp*ED5-1 has been used to make a hybrid protein consisting of the NH_2-terminal part of *trp* D fused to HGH. This was

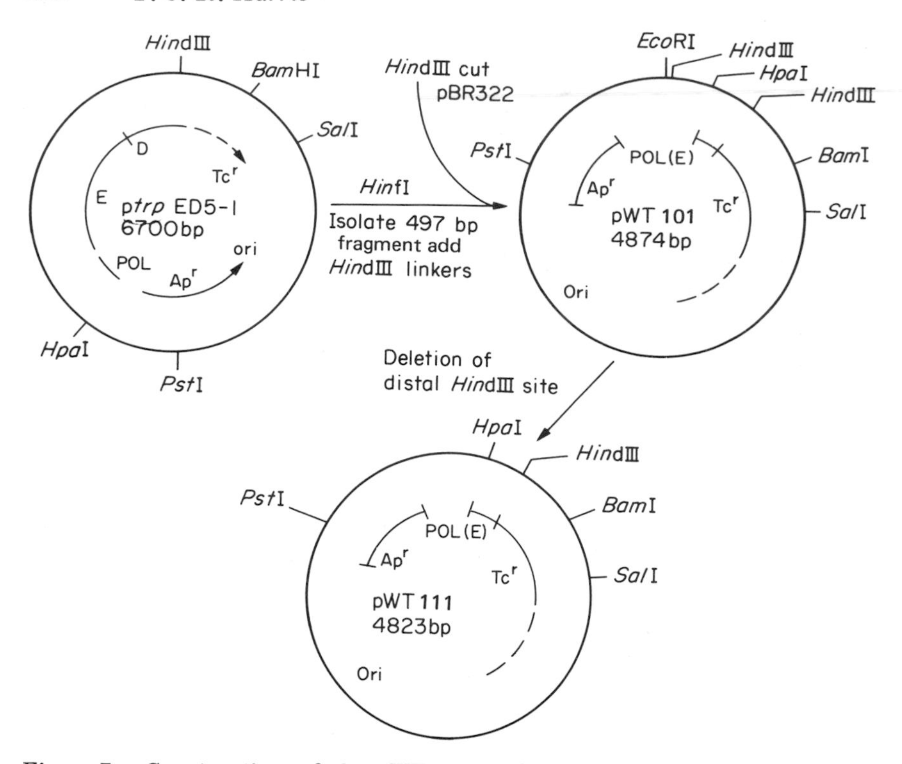

Figure 7 Construction of the pWT series of *trp* promoter expression vectors. Adapted from Hallewell and Emtage 1980, and Tacon *et al.*, 1980.

done by inserting HGH cDNA into the *Hind*III site of the plasmid using *Hind*III linkers maintaining the frame of translation (Martial *et al.*, 1979). About 3% of total cell protein was identified as the hybrid protein by polyacrylamide gel electrophoresis. It is interesting to note that whereas in the original plasmid equal amounts of *trp* E and *trp* D were produced, in the HGH construction the amount of *trp* E made was far greater than the *trp* D fusion protein. This indicates that some feature of the gene other than its transcription or the structure of the rbs affects the overall level of expression. Derivatives of p*trp*ED5-1 have been made by cloning a *Hinf*I fragment, containing the *trp* operator/promoter, into *Hind*III-digested pBR322 using *Hind*III linkers (Tacon *et al.*, 1980). The resulting plasmid (pWT 111, Fig. 7) contained the *trp* regulatory regions plus the first seven codons of *trp* E, upstream of the tetracycline resistance gene of pBR322. Insertion of "foreign" DNA into the *Hind*III site in the correct frame for translation allowed the synthesis of N-terminal *trp* E fusions. Further derivatives, allowing expression of a fusion protein in the other two translation frames, were also constructed using a technique similar to the one used to phase β-galactosidase fusions (Charnay *et al.*, 1979). The tetracycline resistance genes of all these plasmids were placed under *trp* control by the insertion of the *trp* promoter/operator at the *Hind*III site, allowing transcription through any inserted sequence to be monitored by tetracycline resistance.

One of these derivatives (pWT121) has been used to express the haemagglutinin gene (HA) of fowl plague influenza virus (Emtage *et al.*, 1980). Nucleotide sequencing indicated that the frame of translation of *trp* E would be maintained if the HA gene (obtained by cDNA cloning) was inserted into the *Hind*III site of pWT121. The predicted fusion protein consisted of 7 amino acids of *trp* E, 6 amino acids from the linker, 6 phenylalanine residues from the $(T)_{19}$ "tail" on the cloned HA DNA, 7 amino acids from the normally untranslated 5′ end of the HA gene, 558 amino acids of the HA and its signal sequence plus 5 amino acids from the C terminal *Hind*III linker, giving a total of 589 amino acids (Emtage *et al.*, 1980). A protein with HA immunoreactivity was detected in *E. coli* containing the plasmid but the size of the immunoprecipitated polypeptide on polyacrylamide gels was smaller than the predicted 69 000 molecular weight indicating that some processing had occurred (Emtage *et al.*, 1980). The levels of expression were considerably lower than expected compared to the levels of *trp* E produced by p *trp* ED5-1 (Hallewell and Emtage, 1979) but similar to the levels of ovalbumin-like proteins produced as β-galactosidase fusions from the *lac* promoter (Mercereau-Puijalon *et al.*, 1978; Fraser and Bruce, 1978). A possible explanation for the low level of expression was that the eukaryotic signal sequence contained in the HA gene was somehow inhibitory or toxic to *E. coli.* Rose and Shafferman (1981) showed that analogous *trp* expression plasmids containing the complete glycoprotein gene (G) of vesicular stomatitis virus (VSV) (obtained from cDNA copies of virus mRNA) including the signal peptide, fused in phase to the first seven amino acids of *trp* E, were lethal in cells lacking *trp* repressor (*trp* R^-) but could be transformed into *trp* R^+ cells. A protein of the size expected for a *trp* E-G fusion protein was detected in ^{35}S methionine pulse-labelled *trp* R^+ mini cells (where only protein synthesis from plasmid DNA is examined). Deletion of some of the hydrophobic residues at the COOH-terminus of the glycoprotein gene did not overcome the lethality effect in *trp* R^- cells, whereas expression of a gene lacking ten hydrophobic amino acids from the NH_2-terminal signal sequence was detected (Rose and Shafferman, 1981). In neither instance, however, was the level of expression high enough for a new protein to be detected as a stained or labelled band in polyacrylamide gels of total cell protein. Heiland and Gething (1981) similarly obtained low HA activity in *E. coli* from plasmids containing the *lac* promoter directing transcription of short β-galactosidase-HA fusions with a deleted HA signal sequence.

Higher levels of expression of HA (5–7% total cell protein) have been achieved by using longer *lac* promoter β-galactosidase fusions. Fragments of HA, lacking the N-terminal signal sequence were fused to a long NH_2-terminal β-galactosidase coding sequence giving rise to very large (> 130 000 molecular weight) insoluble fusion proteins (Davis *et al.*, 1981). Some expression was also detected (by polyacrylamide gel electrophoresis) when mature HA was fused to a *trp* LE gene (see below) transcribed from the *trp* promoter, but constructions designed to obtain direct expression of mature HA from either the *lac* or *trp* promoter were not successful (Davis *et al.*, 1981).

Other animal virus proteins have been expressed in *E. coli* as *trp* E fusions. Kleid *et al.* (1981a) constructed a vector containing the *trp* promoter/operator, designed to direct the synthesis of the immunogenic capsid protein VPI (also called VP3) of foot and mouth disease virus (FMDV) linked to a hybrid protein, consisting of the NH_2-terminus of the *trp* leader peptide fused to the last third of the *trp* E protein gene (ΔLE 1413). The *trp* LE fragment of 190 amino acids is a particularly useful protein for fusions because it is insoluble and resistant to proteolytic cleavage. Moreover, the deletion (ΔLE 1413) removes the attenuator site so that any mRNA secondary structure effects leading to attenuation should be avoided. (For a detailed discussion of attenuation in the *trp* operon, see Yanofsky, 1981.) The expression vector (pFM1) was made by annealing a fragment of the VPI gene (a *Pst*I-*Pvu*II fragment containing codons 8-211) to the *Eco*RI-*Bam*HI fragment of pBR322 and ligating these to a *trp* promoter/operator fragment containing the *trp* LE hybrid coding sequence. The *trp* fragment contained an *Eco*RI site at the end of the *trp* E gene, so an *Eco*RI-*Pst*I linker was used to link the sequences and maintain the frame of translation (Kleid *et al.*, 1981a). (The details of the construction of this *trp* LE fragment are described in Kleid *et al.*, 1981b.)

A *trp* LE-VPI fusion protein which has a molecular weight of about 44 000 was obtained in the pellet from a detergent lysate of bacteria harbouring the plasmid after low speed centrifugation. Relatively large amounts of the protein (170 mg from 800 ml of culture, or about 17% of total cell protein) were produced. After purification by gel electrophoresis the protein was mixed with adjuvant and used to vaccinate both swine and cattle. As might be expected, the fusion protein was no better as an immunogen than

VPI isolated from the virus particle (Kleid *et al.*, 1981). Nevertheless, the results are of importance because a very high level of expression was obtained and the hybrid protein could be purified sufficiently for it to be used as an immunogen. This will enable studies of antigen presentation and the effect of adjuvants on the immunogenicity of the protein to be assessed properly with reasonable amounts of pure material. Kupper *et al.* (1981) have also described the expression of a VPI fusion protein, using cDNA made to a different FMD virus serotype, coupled to a P_L promoter based expression vector (pPLc24, see Fig. 6). VPI was produced with 99 amino acids of MS2 polymerase at the NH_2-terminus and 13 amino acids from the vector at the COOH-terminus. This construction did not lead to the levels of fusion protein obtained by Kleid *et al.* (1981a).

C Expression of interferon

Plasmids have been constructed containing a leukocyte interferon gene (LeIF-A) downstream of the *trp* promoter and *trp* LE fragment (Goeddel *et al.*, 1980a). Although the precursor form of the protein containing the signal sequence (pre LeIF-A) was fused to the *trp* E gene in phase, the size of the expressed protein was consistent with translation having initiated at the signal peptide AUG codon. Not surprisingly, considerably higher amounts of interferon (480 000 u/l) were obtained in this way, than the 20 000 u/l reported for a pre LeIF D gene cloned directly into the *Pst*I site of pBR322 (Nagata *et al.*, 1980a, see below). The designed expression of native, mature LeIF has also been accomplished using the *trp* promoter. The procedure followed was a variation of the method used to express human growth hormone (Goeddel *et al.*, 1979b). As shown in Fig. 8, the *Sau*3A site between codons 1 and 2 of LeIF-A cDNA was used to produce a fragment (*Sau*3A-*Ava*II) to which a synthetic DNA sequence, coding for an initiating ATG plus a codon for the N-terminal cysteine of mature LeIF and an *Eco*RI cohesive end, was ligated. This reconstituted N-terminal coding sequence was then annealed to the rest of the coding sequence and the gene inserted into a plasmid containing the *trp* promoter/operator and leader peptide rbs upstream of an *Eco*RI site. High levels of expression of LeIF were obtained as measured by the levels of antiviral activity present in bacteria containing this plasmid (2.5×10^8 u/l; 600 μg/l). The protein was soluble and had all the expected characteristics of mature leukocyte interferon (e.g. it was stable at pH 2 and the activity was neutralized by antihuman leukocyte interferon antibodies). Moreover, the protein could be shown to protect monkeys against potentially lethal encephalomyocarditis (EMC) virus infection

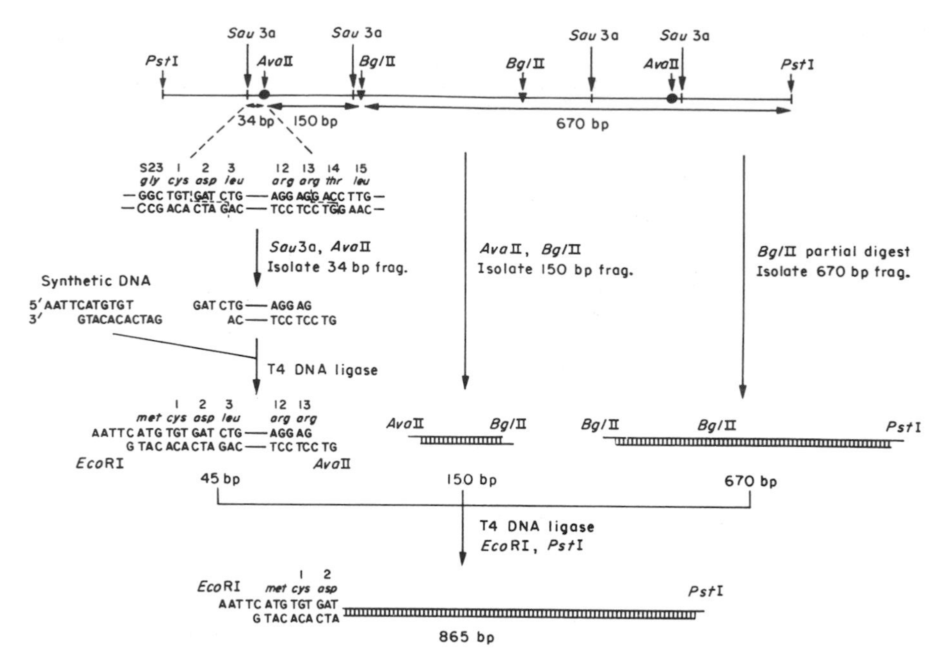

Figure 8 Construction of a gene coding for mature human LeIF-A for expression in a *trp* promoter based vector. From Goeddel *et al.*, 1980a. Reprinted by permission from *Nature*, **287**, 411–416. Copyright 1980, Macmillan Journals Ltd.

even in relatively impure form (Goeddel *et al.*, 1980a). This protein has now been purified from *E. coli* and some of its structural properties examined (Wetzel, 1981; Wetzel *et al.*, 1981b).

Human fibroblast interferon has also been synthesised in *E. coli* as a native, mature protein lacking the signal sequence, using both the *trp* and the *lac* promoter (Goeddel *et al.*, 1980b). The construction of these vectors was more complicated than the similar LeIF vectors described above, owing to the lack of conveniently placed restriction sites. The signal peptide region of pFIF3, a plasmid containing HFIF cDNA, was removed using a modification of the primer repair method described for removing the signal peptide from the influenza HA gene (Davis *et al.*, 1981; see also Kleid *et al.*, 1981b). A HhaI fragment of DNA containing the entire HFIF cDNA sequence was denatured (instead of making DNA with long single stranded 3′ extensions with λ exonuclease -- Davis *et al.*, 1981) and a primer, containing an ATG and the coding sequence for the first four amino acids of mature HFIF, annealed. The Klenow fragment of DNA polymerase was then used for repair synthesis. At the same time the 3′—5′ exonuclease activity of the enzyme removed

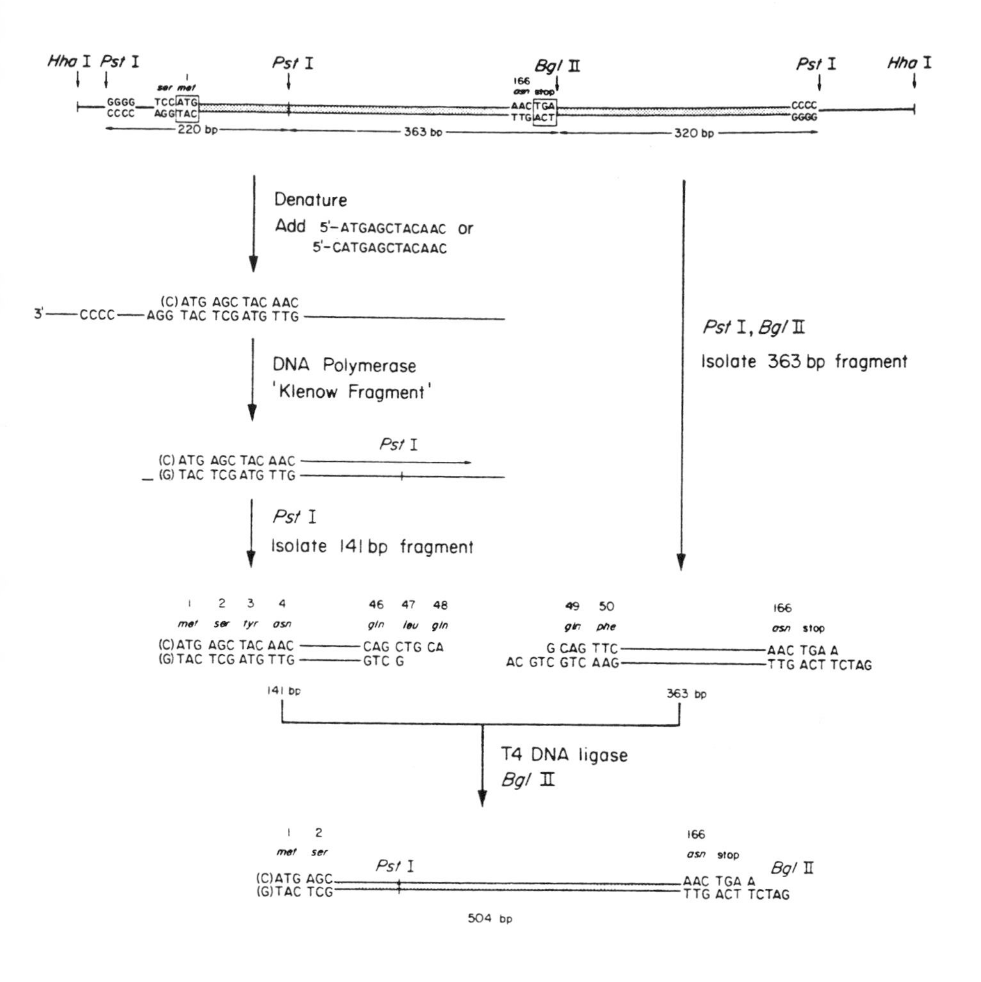

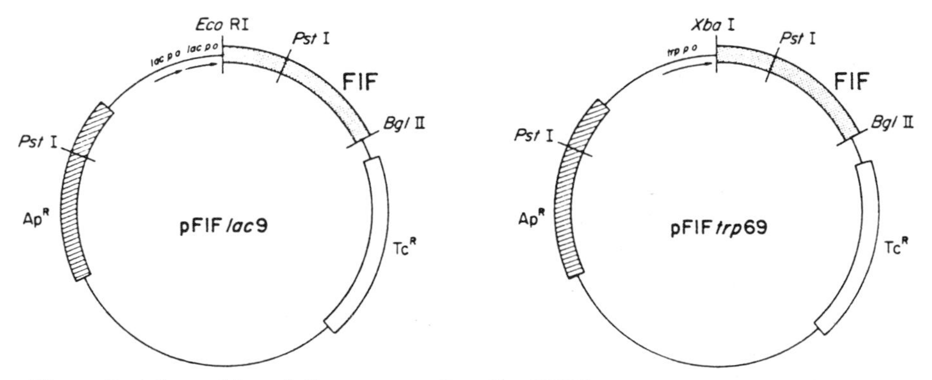

Figure 9 Assembly of the gene coding for HFIF and construction of the *lac* and *trp* based vectors used for expression of the mature protein. Taken from Goeddel *et al.*, 1980b. (Reprinted from *Nucleic Acids Res.* with permission.)

protruding 3′ ends leaving a blunt end where the primer was annealed (Fig. 9). This 5′ fragment was then obtained by restriction enzyme digestion and gel electrophoresis and the coding sequence reestablished by ligation of the fragment to other suitable restriction fragments (Fig. 9). The reconstituted gene was inserted into expression plasmids containing either the *trp* promoter and leader rbs or the *lac* promoter and β-galactosidase rbs (Goeddel *et al.*, 1980b). Bacteria containing the *trp* promoter plasmid apparently produced more HFIF than bacteria containing the *lac* promoter; moreover plasmids containing three *trp* promoters in series before the HFIF gene produced 4—5 times as much HFIF as those plasmids with only one *trp* promoter. Nevertheless, the levels of HFIF produced from one *trp* promoter were 10-fold less than the analogous LeIF constructions (cf. Geoddel *et al.*, 1980a).

Increased levels of synthesis of both LeIF and HFIF have since been obtained by manipulation of the nucleotide sequence between the SD sequence of the *trp* leader and the ATG of the interferon coding sequences (Shepard *et al.*, 1982). The optimal spacing for expression of both types of interferon was nine nucleotides, despite the differences in nucleotide sequence between the SD sequence and the AUG for each transcript. Once again, it is likely that the increase in expression is due to the mRNA having a secondary structure which facilitates ribosome binding and initiation (Shepard *et al.*, 1982; Iserentant and Fiers, 1980; Gheysen *et al.*, 1982). This provides a further indication that the efficiency of translation may be one of the major limiting steps in the production of proteins from cloned eukaryotic genes.

The availability of cloned cDNA copies to several of the different types of leukocyte interferon genes which make up the multigene family (Nagata *et al.*, 1980b, Goeddel *et al.*, 1981) has facilitated the construction of vectors for the expression of other leukocyte interferon genes and has enabled the activities of the expressed proteins to be compared to natural human leukocyte interferon (e.g. see Stewart *et al.*, 1980). Yelverton *et al.* (1981) synthesized LeIF-B by inserting the cDNA for this protein in place of the LeIF-A sequence in the plasmid pLeIFA-25 (Goeddel *et al.*, 1980a, see above). The protein produced had markedly different antiviral specificity to LeIF-A, in agreement with other studies using LeIF molecules obtained from different cDNA clones (Streuli *et al.*, 1980). The fact that the LeIF gene family is unusual in not having introns interrupting the coding sequence (Nagata *et al.*, 1980b) extends the possibilities as far as expression of these genes is concerned, since DNA from genomic clones could be used for constructions, instead of relying on the availability of cDNA clones (e.g. Mory *et al.*, 1981). The observed difference in the biological properties of these *E. coli* derived leukocyte interferons has led to the construction of hybrid genes, consisting of the coding sequence for the NH_2-terminus of one type of interferon (e.g. LeIF A or D) coupled to that coding for the COOH-terminus of another, by recombination *in vitro* through common restriction enzyme sites within their coding regions (Streuli *et al.*, 1981). These hybrid genes have been expressed as short β-galactosidase fusion proteins from the *lac* promoter or as native polypeptides under *trp* promoter control, as described above. Polypeptides with biological characteristics different to either of the parent molecules were obtained (Streuli *et al.*, 1981; Weck *et al.*, 1981).

A potentially greater number of different interferon genes could be constructed synthetically. The availability of a chemically synthesized interferon gene, made from oligonucleotide blocks (Edge *et al.*, 1980) opens up the possibility of altering the coding region specifically by substituting different oligonucleotide blocks. The expression of this chemically synthesised LeIF gene in *E. coli* from the *lac* promoter has been reported recently (De Maeyer *et al.*, 1982). Human immune interferon (interferon γ) has also been cloned, sequenced and the protein expressed to a low level in *E. coli* (Gray *et al.*, 1982), so it should now be possible to compare the specific activity and biological properties of the three different types of human interferon synthesized by *E. coli.*

VIII The β-lactamase promoter

A Synthesis of fusion proteins

An alternative system that has been used to obtain low level expression of hybrid proteins is the β-lactamase (ampicillin resistance) gene of pBR322. The sequence of this gene is known (Sutcliffe, 1979); it has a single *Pst*I site located between codons 181 and 182. The expression studies arose as a result of the fact that the *Pst*I site has been used extensively as the insertion site for cloning cDNA by homopolymer tailing with terminal transferase. Not only does cloning into this site inactivate β-lactamase but the G-C homopolymer tails reconstitute *Pst*I sites at each end of the insert, making removal of the insert potentially easy. (For a detailed description of cDNA cloning see Craig and Hall, 1983.)

Villa-Komaroff *et al.* (1978) cloned cDNA transcripts of rat preproinsulin mRNA into the *Pst*I site of pBR322 by this method. Since G tails of different lengths were added to the plasmid by terminal transferase and the insertion took place within the coding region of

β-lactamase some of the insertions were expected to maintain the reading frame across the G-C junction and give rise to β-lactamase fusion proteins. Using a solid phase radioimmunoassay designed to detect bacterial colonies expressing antigens (Broome and Gilbert, 1978), one of the clones containing insulin sequences was found to produce a β-lactamase-insulin hybrid protein. Furthermore, the hybrid protein was secreted into the periplasmic space indicating that the signal sequence at the NH_2-terminus of β-lactamase was directing secretion of the hybrid protein. Nucleotide sequencing showed that 18 G residues separated the β-lactamase and the rat proinsulin sequence, indicating that a run of six glycines fused the alanine at position 182 of β-lactamase to the glutamine at position 4 of rat proinsulin (Villa-Komaroff *et al.*, 1978).

Rat pregrowth hormone has also been expressed as a β-lactamase hybrid protein by fusion of the β-lactamase gene to the conveniently placed *Pst*I site at codon -24 of the growth hormone signal sequence (Seeburg *et al.*, 1978). Although expression of the hybrid protein, consisting of 181 amino acids of β-lactamase linked to 214 amino acids of rat pregrowth hormone, could be detected in mini-cells, there was little if any growth hormone-related protein detected in the periplasmic space, in contrast to the findings with proinsulin (Villa-Komaroff *et al.* 1978).

B Secretion of native proteins using β-lactamase fusions

The successful transport of a β-lactamase-insulin hybrid to the periplasmic space prompted further experiments to characterise the requirements for the secretion. A series of plasmids (pKT) were made in which the *Pst*I site in the β-lactamase gene was moved so that it was in or near the coding region for the signal sequence of β-lactamase. This was done by deletion mutagenesis using *Bal*31 exonuclease and *Pst*I sites recreated at the boundaries by ligation of *Pst*I linkers. The genes for rat proinsulin and preproinsulin were cloned into the *Pst*I site of these different plasmids, resulting in the formation of different hybrid β-lactamase and insulin signal sequences (Talmadge *et al.*, 1980a). The level of insulin accumulating in the periplasmic space of *E. coli* containing the different plasmids was then measured. It was concluded that possession of a signal sequence was essential for secretion and that this sequence could be either prokaryotic or eukaryotic (or both) in origin. In addition it was shown that for constructions where the entire rat preproinsulin sequence was fused to all or part of the bacterial signal sequence, proinsulin could be detected in the periplasmic space, indicating that the bacterial signal peptidase was recognising the eukaryotic signal sequence and processing preproinsulin to proinsulin (Talmadge *et al.*, 1980b). Similar results have been reported for human preproinsulin attached to a hybrid signal sequence (Chan *et al.*, 1981).

More recently, Talmadge *et al.* (1981) have shown that if the rat preproinsulin sequence was fused to the first eight amino acids of β-galactosidase under *lac* promoter control, then the expressed protein was processed at the eukaryotic signal sequence as before, and proinsulin found in the periplasmic space. This finding provides additional evidence that it is the signal sequence that is responsible for the secretion, rather than any bacterially derived peptide remaining at the NH_2-terminus. It also demonstrated that the signal sequence need not be precisely at the NH_2-terminus; in these β-galactosidase fusions, 18 amino acids preceded the preproinsulin sequence (Talmadge *et al.*, 1981). One of the potential advantages of making DNA constructions that ensure the secretion of the protein in this way is that periplasmic proteins may be less likely to be degraded than those remaining in the cytoplasm. In this regard, Talmadge and Gilbert (1982) have shown that the rat proinsulin remaining in the cytoplasm produced from constructions not having a signal sequence is more rapidly degraded than proinsulin that is secreted. Whether this will be true for other eukaryotic proteins remains to be seen but it should be noted that similar experiments with other eukaryotic genes have not been reported.

C Synthesis of other native proteins

Mouse dihydrofolate reductase (DHFR) cDNA has also been inserted into the *Pst*I site of pBR322 by G-C tailing and enzymatically active protein detected (Chang *et al.*, 1978). Phenotypic selection for expression of the eukaryotic sequence (i.e. resistance to trimethoprim) allowed the isolation of bacterial cells making a soluble protein that had the enzymatic and immunological properties and molecular size (22 000 molecular weight) of the native mouse enzyme. Nucleotide sequence analysis across the *Pst*I site of the plasmid-cDNA junctions showed that the DHFR cDNA inserts were not in the correct reading frame for expression to be occurring as a fusion polypeptide, implying that initiation of translation was occurring at the DHFR AUG in a polycistronic mRNA. Moreover, some expression was detected in clones with inserts in the opposite orientation to that required for a fusion polypeptide (Chang *et al.*, 1978). DHFR cDNA clones with different sequences before the ATG induced different levels of trimethoprim resistance, indicating that the amount of DHFR made was being controlled at the translational level (Chang *et al.*, 1978). It was postulated that the *Pst*I-poly(G) sequence at the vector/cDNA junction provided a sufficient homology with the SD sequence for it to direct ribosomes to initiate translation at a nearby AUG. Further evidence for this idea has come from a more detailed correlation of the nucleotide sequences at the vector-insert junction with the levels of enzyme made, for several of the expressing clones (Chang *et al.*, 1980). It was apparent that the optimal distance between the beginning of the artificial SD sequence and the AUG was 12—14 nucleotides — somewhat longer than the optimum found for other genes (e.g. interferon, see Shepard *et al.*, 1982). This difference may be attributable to mRNA secondary structure and the degree of homology between the SD sequence and the 3′-end of 16S rRNA (Chang *et al.*, 1980; Iserentant and Fiers, 1980). In this connection Jay *et al.* (1981) have joined a chemically synthesised rbs sequence with maximum homology to the 3′-end of 16S rRNA, to the SV40 tAg coding sequence and placed this hybrid gene into the *Pst*I site in the β-lactamase gene. The synthetic rbs present in the mRNA (transcribed from the β-lactamase promoter) was apparently recognised because authentic SV40 tAg could be detected in bacterial extracts by immunoprecipitation (Jay *et al.*, 1981). Manipulation of the rbs sequence will now allow a more systematic study of the effect of rbs-16S rRNA homology on the translation of cloned eukaryotic genes.

Phenotypic selection has also been used to obtain plasmids expressing the herpes simplex virus (HSV) thymidine kinase (TK) gene (Garapin *et al.*, 1981). A fragment of the virus DNA containing the gene was fused to each of the three "phased" *lac* promoter β-galactosidase fusion vectors (Charnay *et al.*, 1978) and the plasmids used to transform TK^- *E. coli* to TK^+. When the TK gene was in the right orientation for transcription from the *lac* promoter functional thymidine kinase was found with all three vectors, irrespective of the translational reading frame. Analysis of the nucleotide sequence preceding the start sites identified putative SD sequences, suggesting that initiation of translation was occurring as described for DHFR (Chang *et al.*, 1980). Native thymidine kinase has also been detected in bacteria harbouring plasmids containing the HSV TK gene positioned downstream of the tetracycline promoter of pBR322 (Kit *et al.*, 1981). The plasmids constructed by Nagata *et al.* (1980a) producing LeIF as a β-lactamase fusion using the pKT vectors also appeared to give rise to non-fusion polypeptides, since their production was similarly independent of the translational reading frame predicted by the vector into which the cDNA was inserted.

Reinitiation of translation has also been invoked to explain the production of hepatitis virus core antigen (HBcAg) from restriction fragments of hepatitis virus DNA cloned into the *Pst*I site of pBR322 by G-C tailing (Burrell *et al.*, 1979; Pasek *et al.*, 1979). Although all the inserts were in the correct orientation for expression of β-lactamase fusions, none of them were in the right translation frame for the core antigen to be made as a β-lactamase fusion polypeptide, indicating that initiation was occurring from an ATG in HBcAg. Sufficient core antigen was produced in this way for it to be used to raise a specific antiserum (Pasek *et al.*, 1979). Higher levels of HBcAg have now been expressed in *E. coli* from both the *trp* and *lac* promoters as native and fusion polypeptides (Edman *et al.*, 1981; Stahl *et al.*, 1982). For the expression of the native polypeptide a vector was constructed that contained the *trp* promoter and leader rbs abutted to the HBcAg coding sequence, giving an SD to ATG distance of 13—16 base pairs (Edman *et al.*, 1980). Under conditions of full derepression, 10% of newly synthesized protein was HBcAg. The HBcAg contains 25 arginine codons, 17 of which are coded for by a codon rarely used in *E. coli* (Grantham *et al.*, 1980) so the high level of synthesis obtained suggests that, for this gene at least, the concentration of iso-accepting tRNA species does not inhibit translation. Gel exclusion chromatography and electron microscopy of bacterial HBcAg produced by *lac* promoter constructions (Stahl *et al.*, 1982) has shown that the antigen is in an aggregated form similar in appearance to virus cores seen in extracts of HBV infected liver cells (Cohen and Richmond, 1982). The aggregated protein seems to recognise HBcAg antibodies well enough for it to be used as a diagnostic reagent (Stahl *et al.*, 1982).

Small amounts of another hepatitis B virus antigen, the surface antigen (HBsAg) have also been synthesized in *E. coli* from virus DNA cloned into the *Pst*I site of pBR322 (Pasek *et al.*, 1979; Mackay *et al.*, 1981). Attempts to make larger amounts of this protein as a native polypeptide from the *trp* promoter have not been successful, although some expression of a protein lacking the HBsAg signal sequence but fused to the NH_2-terminus of the signal sequence of β-lactamase transcribed from the *trp* promoter, has been reported (Edman *et al.*, 1981). Very small amounts of long β-galactosidase-HBsAg fusion proteins have also been detected in *E. coli* infected with a recombinant λ phage carrying the HBsAg gene and the *lac* promoter (λ p *lac* 5-1 UV5, Charnay *et al.*, 1980).

IX Conclusions and future prospects

Table 2 presents an up to date summary of the higher eukaryotic proteins that have been expressed in *E. coli*. The levels of expression are indicated where possible and the nature of the construction

Table 2 Expression of eukaryotic DNA in *E. coli.*

Protein (molecular weight)	Source of DNA	Promoter	Construction	Nature of product/cleavage protocol	Cell fraction (before or after lysis)	Level of expression (% total cell protein or mol/cell)	Reference
Somatostatin (c. 1500)	Chemical synthesis	*lac*	β-gal fusion	Hybrid protein; CNBr cleavage	Pellet	Low but immuno-detectable c 0.05%	Itakura *et al.*, 1977
Human insulin A = (c. 2000) B = (c. 3000)	Chemical synthesis of separate A and B chains	*lac*	β-gal fusion	Hybrid A and B CNBr cleavage. Re-constitution to give insulin	Pellet	20% = 10^5 mol/cell; 10 mg/24 g wet cells	Goeddel *et al.*, 1979a
Human proinsulin (mini C) (c. 6500)	Chemical synthesis	*lac*	β-gal fusion	Hybrid protein; CNBr cleavage	Pellet	12 mg/322 g wet cells	Wetzel *et al.*, 1981a
Rat proinsulin (c. 12 000)	cDNA	*lac*	β-gal fusion	Proinsulin	Periplasmic space	Low	Talmadge *et al.*, 1981
Thymosin α_1 (3100)	Chemical synthesis	*lac*	β-gal fusion	Hybrid protein; Nα desacetyl thymosin α_1 formed after CNBr cleavage	Pellet	10^5 mol/cell 40 mg/100 g wet cells	Wetzel *et al.*, 1980
Leu-enkephalin (c. 600)	Chemical synthesis	*lac*	β-gal fusion	Hybrid protein; CNBr gives native hormone	Pellet	2×10^5 mol/cell 5 mg/100 g wet cells	Shemyakin *et al.*, 1980
α neo-endorphin (c. 1200)	Chemical synthesis	*lac*	β-gal fusion	Hybrid protein; CNBr gives native hormone	Pellet	5×10^5 mol/cell 4 mg/10.9 g wet cells	Tanaka *et al.*, 1982
β-endorphin (c. 3200)	ACTH/ βLPH cDNA	*lac*	β-gal fusion	Hybrid protein; Native hormone prepared by citraconylation and trypsin treatment	Pellet	5%	Shine *et al.*, 1980

Rous Sarcoma virus (RSV) protein kinase (src) (60 000)	cDNA	*lac* (× 2)	8 amino acid β-gal fusion	Hybrid protein; NH_2 terminal non src amino acids	Supernatant	0.3%	Gilmer *et al.*, 1982; Gilmer and Erikson, 1981
HSV thymidine kinase (TK) (43 000)	Virus DNA clone	*lac*	Short β-gal fusion	Native enzyme? (40—42 000)	Supernatant	0.2—0.3%	Garapin *et al.*, 1980
Chicken ovalbumin (45 000)	cDNA	*lac*	Short β-gal fusion	β-gal-ovalbumin	Pellet (some in periplasmic space)	1—1.5% 5×10^4 mol/cell	Fraser and Bruce, 1978; Mercereau-Puijalon *et al.*, 1978; Baty *et al.*, 1981
Rabbit β globin (c. 9000)	cDNA	*lac*	rbs fusion	Met-β globin	Pellet?	1.5×10^4 mol/cell	Guarente *et al.*, 1980
Human growth hormone (HGH) (24 000)	Chemical synthesis and cDNA	*lac* (× 2) and *lac-trp* (*tac*)	Constructed rbs fusion, SD-AUG is 7—11 bp	Met HGH	Supernatant (soluble protein)	1.8×10^5 mol/cell 2.4 μg/ml	Goeddel *et al.*, 1979b de Boer *et al.*, 1982
Human fibroblast interferon (HFIF) (23 000 — Pre-HFIF)	cDNA	*lac*	rbs fusion	Pre-HFIF Met-HFIF	Pellet (protein aggregates)	Low (50 mol/cell)	Taniguchi *et al.*, 1980
(20 000 — HFIF)	cDNA	*lac*	rbs fusion	Met-HFIF	Supernatant	2.2×10^3 mol/cell	Goeddel *et al.*, 1980b
Human leukocyte interferon (LeIF-α) (20 000)	Chemical synthesis	*lac*	rbs fusion	Met-LeIF	Supernatant	c. 10^7 u/ml	De Maeyer *et al.*, 1982
LeIF-α_2	cDNA	*lac* (in M13 mp7)	Short β-gal fusion	Native LeIF-α_2 19 500	Supernatant	10^9 u/l = 5 mg/l	Slocombe *et al.*, 1982

Thaumatin (22 000)	cDNA	*lac*	rbs fusion	Preprothaumatin	—	Low	Edens *et al.*, 1982
Influenza virus HA gene (61 000)	cDNA	*lac*	linker-part HA fusion	HA protein lacking some NH_2-terminal amino acids	—	3×10^3 mol/cell	Heiland and Gething, 1981
			Long β-gal fusion	Large hybrid protein (130 000)	Pellet	5—7% 5—7 × 10^4 mol/cell	Davis *et al.*, 1981
Hepatitis B virus core antigen (HBcAg) (19 000 or 22 000)	Virus DNA clone	*lac*	short β-gal fusion	Native and hybrid HBcAg	Supernatant (small aggregate)	Low	Stahl *et al.*, 1981
Hepatitis B virus surface antigen (HBsAg) (22 600)	Virus DNA clone	*lac* (in phage)	long β-gal fusion	Hybrid protein (138 000)	Supernatant	0.05%	Charnay *et al.*, 1980
Polyoma small t antigen (20 000)	Virus DNA	*lac*	β-gal linker fusion	Hybrid protein (26 000)	Supernatant (soluble)	0.15%	Horwich *et al.*, 1980
SV40 tAg (19 000)	Virus DNA	*lac*	rbs fusion	Native tAg	Pellet; high mol wt complex	1—5 × 10^3 mol/cell 0.8%, 4—8 × 10^4 mol/cell	Roberts *et al.*, 1979c; Thummel *et al.*, 1981
SV40 tAg (19 000)	Virus DNA	P_L^*	rbs fusion	Native protein (19 000 and 14 500 derivative	Pellet	2.5% 7%	Derom *et al.*, 1982
HFIF (20 000)	cDNA	P_L^*	β-lactamase or MS_2 polymerase fusions	Native HFIF by processing?	Pellet	Low	Derynck *et al.*, 1980
Foot and mouth disease virus VP1 (FMDV-VP1) (24 000)	cDNA from virus RNA	P_L^*	MS_2 polymerase fusion	MS_2 pol-VP1 fusion polypeptide	Pellet	10^3 mol/cell	Küpper *et al.*, 1981

HGH (24 000)	cDNA	*trp*	*trp* D fusion	*trp* D-HGH hybrid protein (32 000)	Supernatant	3%	Martial *et al.*, 1979
Influenza HA gene (61 000)	cDNA	*trp*	*trp* E-linker fusion	Hybrid protein (69 000) some processing to (61 000)	Pellet?	0.75%	Emtage *et al.*, 1980
VSV G protein (62 000)	cDNA	*trp*	*trp* E fusion	Hybrid protein (lacking G signal sequence)	Supernatant	1%	Rose and Shafferman, 1981
FMDV-VP1 (24 000)	cDNA	*trp*	*trp* LE fusion	Hybrid protein (44 000)	Pellet	17% $1–2 \times 10^6$ mol/cell	Kleid *et al.*, 1981a
RSV src (60 000)	cDNA	*trp*	Short HGH-NH_2-terminal fusion	Hybrid protein (60 000)	Supernatant	5% 1.7×10^5 mol/cell	McGrath and Levinson, 1982
HBcAg (22 000)	Virus DNA clone	*trp*	*trp* leader rbs fusion	HBcAg	100S complex	10% 5×10^5 mol/cell	Edman *et al.*, 1981
HBsAg (22 600)	Virus DNA clone	*trp*	β-lactamase fusion	Hybrid protein (41 000)	—	8.5% 1.7×10^5 mol/cell	Edman *et al.*, 1981
LeIF-A (c. 20 000)	cDNA	*trp*	*trp* LE fusion	Pre LeIF-A	Supernatant	1 μg/l	Goeddel *et al.*, 1980a;
			trp leader rbs fusion	Met LeIF-A	Supernatant	600 μg/l	Shepard *et al.*, 1982
Le-IF B (20 000)	cDNA	*trp*	*trp* leader rbs fusion	Met LeIF-B	Supernatant	200 μg/l	Yelverton, *et al.*, 1981
HFIF (20 000)	cDNA	*trp* (× 3)	*trp* leader rbs fusion	Met HFIF	Supernatant	$4.5–20 \times 10^3$ mol/cell	Goeddel *et al.*, 1980b
Human immune interferon (17 000?)	cDNA	*trp*	*trp* leader rbs fusion	Met IF-γ	Supernatant after sonication	Low c. 80 mol/cell	Gray *et al.*, 1982 1982

Human serum albumin (67 000)	cDNA	*trp*	*trp* leader rbs fusion	Met-HSA	—	Low	Lawn *et al.*, 1981
Chymosin (35 000)	cDNA	*trp*	*trp* leader rbs fusion	Prochymosin	Pellet	1–5% 5×10^4 mol/cell	Emtage *et al.*, 1983
Thaumatin (22 000)	cDNA	*trp*	*trp* leader rbs fusion	Preprothaumatin	—	Low	Edens *et al.*, 1982
Rat prepro insulin (12 000)	cDNA	β-lactamase	β-lactamase fusion	Hybrid protein	Periplasmic space	100 mol/cell	Villa-Komaroff *et al.*, 1978
			β-lactamase/ preproinsulin signal sequence fusions	Proinsulin	Periplasmic space	6×10^3 mol/cell	Talmadge *et al.*, 1980b; Talmadge and Gilbert, 1982
Human prepro-insulin (12 000)	cDNA	β-lactamase	β-lactamase fusions	β-lactamase/ preproinsulin hybrid proteins and proinsulin	Periplasmic space	0.2–1%	Chan *et al.*, 1981
Rat prepro-hormone (c. 22 000)	cDNA	β-lactamase	β-lactamase fusion	Hybrid protein (44 000)	—	2.4×10^4 mol/cell	Seeburg *et al.*, 1978
Rat prolactin (22 000)	cDNA	β-lactamase	β-lactamase fusion	Hybrid protein	—	Low	Erwin *et al.*, 1981
Bovine growth hormone (21 000)	cDNA	β-lactamase	β-lactamase fusion	Hybrid protein	—	Low	Keshet *et al.*, 1981
Porcine secretion (c. 3000)	Chemical synthesis	β-lactamase	β-lactamase fusion	Pre β-lactamase and β-lactamase-secretin hybrids (23 000 and 19 000)	Supernatant	2 mol/cell (desamido secretin)	Suzuki *et al.*, 1982

LeIF (20 000)	cDNA	β-lactamase	Short β-lactamase fusions	Met LeIF	Supernatant	1–2 mol/cell	Nagata *et al.*, 1980; Streuli *et al.*, 1980
Urokinase (35 000)	cDNA	β-lactamase	β-lactamase fusion	Enzymatically active proteins (hybrids?) (32 000–150 000)	Supernatant (periplasmic space)	Low	Ratzkin *et al.*, 1981
Mouse DHFR (22 000)	cDNA	β-lactamase	β-lactamase fusion	Met-DHFR	—	Low	Chang *et al.*, 1978; 1980
SV40 tAg (19 000)	Virus DNA clone	β-lactamase	Synthetic rbs fusion within β-lactamase gene	SV40 tAg	—	0.4%	Jay *et al.*, 1981
HBcAg (22 000)	Virus DNA clone	β-lactamase	β-lactamase fusion	HBcAg (native?)	—	Low	Burrell *et al.*, 1979; Pasek *et al.*, 1979
HBsAg (22 600)	Virus DNA clone	β-lactamase	Short β-lactamase fusions	HBsAg	Supernatant	Low	Mackay *et al.*, 1981
HSV TK (43 000)	Virus DNA clone	Tet gene (pBR322)	Tet gene fusion, internal rbs	Enzymatically active protein (40 000)	Supernatant	Low	Kit *et al.*, 1981

*Only constructions using P_L on a plasmid are considered.

(i.e. either rbs or coding sequence fusion) and of the polypeptide produced (either native or hybrid) are noted. The promoter from which the mRNA coding for the protein was transcribed is also indicated.

Several interesting points arise from Table 2. Firstly, there are very few reports of eukaryotic proteins expressed to really high levels in *E. coli*. The best examples are insulin, growth hormone, interferons (various) and the *trp* LE FMDV VPI fusion protein. It is significant that these proteins are the furthest advanced towards a clinical application. For example, insulin, prepared by recombinant DNA techniques is already under extensive clinical trial in the UK (Clark *et al.*, 1982). Similarly, interferon and growth hormone (Hintz *et al.*, 1982) are in clinical trials and the FMDV-VPI fusion is being tested as a vaccine (Kleid *et al.*, 1981a). By far the majority of the proteins in Table 2, however, have not been reported to be expressed to the same high levels. The importance of this limitation, as far as commercial considerations are concerned, is clearly related to the value of the product. For a protein such as SV40 tAg, the levels of expression may well be sufficient to enable the protein to be purified and its function and structure analysed (Thummel *et al.*, 1981) but for other proteins (e.g. chymosin, Emtage *et al.*, 1983) higher expression levels may well be necessary.

A Alternative promoters and constructions

The *lac*, *trp* and P_L promoters fulfill several criteria which make them attractive promoters to use to drive transcription of eukaryotic genes in *E. coli*. They are controlled by repressors which can be removed or inactivated when expression is required, and this controllability may be very important when considering proteins which are toxic to the cell (e.g. hydrophobic proteins). Nevertheless, there are several other strong promoters which have not yet been harnessed to express eukaryotic genes. The promoter for the outer membrane lipoprotein of *E. coli* (the most abundant protein in *E. coli*), a constitutive strong promoter, has been utilized to construct versatile expression vectors. The approach has been to make the *lpp* promoter controllable by inserting a *lac* promoter/operator fragment between it and the 5′-end of the lipoprotein gene. The expression of genes placed downstream of the promoters (the cloning of which is facilitated by a linker containing restriction sites) is then repressed in bacterial strains containing the I^q mutation (which overproduce repressor) but can be induced by IPTG. The 3′-end of the *lpp* gene is maintained in the vector to provide both transcriptional and translational stop signals (Nakamura *et al.*, 1982; Nakamura and Inouye,

1982). The presence of both promoters results in high levels of synthesis of lipoprotein and β-galactosidase when these genes are cloned in the vectors (Nakamura *et al.*, 1982).

A hybrid promoter consisting of the −35 region of the *trp* promoter fused to the *lac* −10 region and the *lac* operator has also been constructed. This promoter (the *tac* promoter) which is regulated by the *lac* repressor (from *lac* I^q) but has the strength of the *trp* promoter, has been used for the controlled expression of HGH (de Boer *et al.*, 1982). One of the problems with strong promoters, however, is that they are difficult to clone, presumably owing to transcriptional blocking of the origin of replication and other plasmid functions. The T-odd phages probably contain the strongest promoters known in terms of RNA polymerase complex formation and rate of RNA chain initiation *in vitro*. Attempts to clone the early promoter of T5 phage were unsuccessful until a vector containing the major terminator of phage *fd* was used (Gentz *et al.*, 1981); similar vectors are being used to clone the early promoter of phage T7. The demonstration by McAllister *et al.* (1981) that T7 RNA polymerase (an early gene product) can utilize T7 late promoters in plasmids, with its usual specificity, suggests a further expression system where the coding sequence for a desired protein is cloned behind a promoter for T7 polymerase and the polymerase itself is supplied to the cell by infection or from a cloned polymerase gene (McAllister *et al.*, 1981).

Another two-plasmid expression system is also being developed. Sninsky *et al.* (1981) have constructed vectors for the temperature regulated expression of cloned genes. In this system a plasmid, containing the *lac* repressor, which fails to replicate at 42°C, coexists with a plasmid that undergoes multicopy "runaway" replication at 42°C (Uhlin *et al.*, 1979) and which contains a cloned gene under the control of the *lac* promoter/operator. Concurrent derepression of the *lac* promoter and copy number amplification occur at elevated temperatures leading to extensive transcription of the cloned gene. As a model system, Sninsky *et al.* (1981) reported the high level expression of chloramphenicol transacetylase (the CAT gene product) at 42°C. Relatively high copy number vectors have also been prepared by cloning fragments of λ DNA into pBR322 (Rao and Rogers, 1978). In addition, the copy number of Col E1 has been increased by deletion mutagenesis (Twigg and Sherratt, 1980) and single base pair substitution (Muesing *et al.*, 1981). Whether these higher copy number vectors or the more complicated bipartite expression systems can be used on a large scale remains to be seen. One of the problems in a scaled up system could well be that of plasmid instability. It may be possible to control stability to some extent (in the absence of a selection, such as antibiotic resistance) by insertion of DNA sequences which are involved in plasmid partitioning during growth (e.g. the *par* locus, Meacock and Cohen, 1980) into the vectors.

Another interesting expression system that has been developed recently utilizes the high copy number of the RF (replicative form) of the single stranded DNA phage M13. Slocombe *et al.* (1982) achieved high levels of expression of LeIF $\alpha 2$ (5 mg/l) by cloning LeIF cDNA adjacent to the *lac* promoter in M13 mp7 to make an NH_2-terminal-short β-galactosidase fusion protein.

B mRNA structure and stability

Considerable evidence has accumulated that the secondary structure of the mRNA around the rbs and AUG is important in determining its translational efficiency (Iserentant and Fiers, 1980; Shepard *et al.*, 1982; Gheysen *et al.*, 1982) but there have been few studies on the stability of the analogue of eukaryotic mRNA in *E. coli*. The expression of *N. crassa* dehydroquinase is increased in bacterial strains lacking polynucleotide phosphorylase, apparently owing to the fact that this phosphorylase is involved in the turnover of eukaryotic mRNA and its absence increases the half life of these molecules (Hautala *et al.*, 1979). The use of bacterial strains lacking polynucleotide phosphorylase has since been reported for expression of other proteins such as insulin (Talmadge and Gilbert, 1982) but not to any large extent, possibly because not all mRNAs are stabilised in this way.

C Nature of the proteins produced

Another important point to emerge from Table 2 is the nature of the proteins that have been produced from *E. coli*. A number of hormones and other effector proteins and some virus antigens have been expressed but there are only a few examples of eukaryotic proteins whose primary function is enzymatic, e.g. DHFR (Chang *et al.*, 1978; 1980); urokinase (Ratzkin *et al.*, 1981); Rous sarcoma virus *src* protein kinase (Gilmer and Erikson, 1981; McGrath and Levinson, 1982), HSV thymidine kinase (Garapin *et al.*, 1981; Kit *et al.*, 1981) and calf chymosin (Emtage *et al.*, 1982). The lack of examples of enzymes may simply reflect the fact that fewer experiments have been done with genes coding for eukaryotic enzymes than have been done with hormone genes. Alternatively it may be a reflection of the fact that large proteins do not seem to be made (or tolerated) as well in *E. coli* as smaller ones (c. 20 000 molecular weight, see Table 2); enzymes tend to be larger proteins than hormones.

The problem of degradation of "foreign" proteins in *E. coli* may well be difficult to resolve. Nevertheless, the observation that hybrid signal sequences can be used to transport insulin to the periplasmic space (Talmadge *et al.*, 1980a, b) has given the option of making constructions designed to ensure secretion (Talmadge and Gilbert, 1982). Extension of this approach to other proteins will be useful and vectors based on the *lpp* gene (a secreted protein) have been constructed for this purpose (see Nakamura and Inouye, 1982). The use of bacterial strains lacking one or more of the normal complement of proteases in *E. coli* (e.g. Lon$^-$, Gottesman *et al.*, 1981; for review of *E. coli* proteases see Goldberg *et al.*, 1981) may be another way of preventing or slowing down degradation of eukaryotic proteins.

Another aspect to be considered is that the conformation of the protein (determined by the primary amino acid sequence) may actually not be the same when made in *E. coli* as it is when made in its normal location. This might explain why several normally soluble proteins are found to be insoluble when made in *E. coli* (Table 2). It is interesting that the available evidence for HGH and leukocyte interferon which are both produced as soluble proteins in *E. coli* indicates that these small proteins (c. 20 000 molecular weight) have S-S bridges formed correctly (Kohr *et al.*, 1982; Wetzel, 1981). Lack of the proper S-S bridges or incorrect folding due to cross-bridging might lead to insolubility. Further work is clearly needed towards gaining an understanding of the conformation of eukaryotic proteins during their synthesis by *E. coli.*

D Other host-vector systems

Several other host and vector systems are being developed for the expression of eukaryotic DNA. Detailed discussion of them is outside the scope of this review but it is nevertheless worthwhile mentioning some of the alternatives to *E. coli* that are becoming available. There is considerable interest in developing expression systems in *B. subtilis* largely because this organism is a non-pathogenic, Gram positive bacterium which does not produce endotoxins. Moreover strains of *Bacillus* are used widely commercially and several secrete large amounts of extracellular proteins. So far, however, there are few reports of the successful expression of eukaryotic proteins in *B. subtilis*, even though molecular cloning in this organism is now an established technique (see Gryczan, 1982). As an example, Hardy *et al.* (1981) have recently reported the expression of native HBcAg and an FMDV-VPI fusion polypeptide in *B. subtilis* utilizing an erythromycin resistance gene promoter. The obligate methylotroph,

M. methylotrophus is also being used as an alternative host since it can be fermented on a large scale using cheap feedstock (ammonia and methanol). The vector is a composite plasmid (pGSS15) containing the antibiotic resistance genes of pBR322 and the broad host range characteristics of R300B, an Inc Q group plasmid. DHFR (from the β-lactamase promoter), an ovalbumin fusion protein and interferon (both from the *lac* promoter) have been expressed in this system (Hennam *et al.*, 1982; De Maeyer *et al.*, 1982).

Considerable effort is also going into developing cloning and expression systems in other organisms fermentable on a large scale such as *Streptomyces* (Chater *et al.*, 1982) and the yeast *S. cerevisiae* (Hollenberg, 1982). There are reports of the synthesis of cloned prokaryotic enzymes (Hollenberg, 1982) and low levels of ovalbumin in yeast (Mercereau-Puijalon *et al.*, 1980). More recently leukocyte interferon has been expressed to high levels in yeast, using either the promoter for the alcohol dehydrogenase I gene or the phosphoglycerate kinase gene promoter, in vectors based on the 2 μ circle (Hitzeman *et al.*, 1981; Tuite *et al.*, 1982). However, at the same time, it has been reported that rat growth hormone is not made in yeast using a similar expression vector containing the rat growth hormone gene (quoted in Hitzeman *et al.*, 1981).

Finally, there is increasing interest in using eukaryotic cells to express eukaryotic proteins. This might be considered rather obvious but it is only very recently that effective vectors have become available (mostly based on virus replicons) for eukaryotic cloning (for review, see Rigby, 1982). There is no doubt that these systems will become important for the analysis of the mechanism of eukaryotic gene expression and its control. It will be interesting to see whether the systems are also exploited for expressing cloned genes in the way *E. coli* has been. Some problems such as those of post-translational modification and stability may be diminished by a eukaryotic expression system but this advantage may be offset by, for example, greater difficulty in controlling transcription or lower yields of biomass or of the desired protein. No doubt, a comparison of the expression of cloned eukaryotic genes in *E. coli* with their expression in eukaryotic cells will be the subject of a future review.

X Acknowledgments

I would like to thank Spencer Emtage, Mike Doel, Gwyn Humphreys and particularly Norman Carey for their constructive criticisms of the manuscript and many other colleagues for helpful advice and information. I am also grateful to Margaret Turner and Jackie

Hardwick for typing the manuscript and to Ed Trewhella and Celltech's internal library service for provision of space and computer on-line search time.

XI References

Backman, K. and Ptashne, M. (1978). Maximizing gene expression on a plasmid using recombination *in vitro*. *Cell* **13**, 65–71.

Backman, K., Ptashne, M. and Gilbert, W. (1976). Construction of plasmids carrying the CI gene of bacteriophage λ. *Proc. Natn. Acad. Sci. U.S.A.* **73**, 4174–4178.

Bassford, P., Beckwith, J., Berman, M., Brickman, E., Casadaban, M., Guarente, L., Saint-Girous, I., Sarthy, A., Schwartz, M., Schuman, H. and Silhavy, T. (1978). Genetic fusions of the *lac* operon. A new approach to the study of biological processes. *In* "The Operon" (Miller, J. H. & Reznikoff, W. S., eds), pp. 245–261. Cold Spring Harbor.

Baty, D., Mercereau-Puijalon, O., Perrin, D., Kourilsky, P. L. and Lazdunsky, C. (1981). Secretion into the bacterial periplasmic space of chicken ovalbumin synthesized in *Escherichia coli*. *Gene* **16**, 79–87.

Beggs, J. D. (1982). Gene cloning in yeast. *In* "Genetic Engineering" (Williamson, R., ed). Vol. 2, 175–203. Academic Press, London.

Bernard, H. U., Remaut, E., Hershfield, M. V., Das, H. K., Helinski, D. R., Yanofsky, C. and Franklin, N. (1979). Construction of plasmid cloning vehicles that promote gene expression from the bacteriophage lambda P_L promoter. *Gene* **5**, 59–76.

Bollen, A., Glineur, C., Delcuve, G., Loriau, R. and Herzog, A. (1981). Expression in *Escherichia coli* of urokinase antigenic determinants. *Biochem. Biophys. Res. Comm.* **103**, 391–401.

Breathnach, R. and Chambon, P. (1981). Organisation and expression of eukaryotic split genes coding for proteins. *Ann. Rev. Biochem.* **50**, 349–383.

Broome, S. and Gilbert, W. (1978). Immunological screening method to detect specific translation products. *Proc. Natn. Acad. Sci. U.S.A.* **75**, 2746–2749.

Brousseau, R., Scarpulla, R., Sung, W., Hsiung, H. M., Narang, S. A. and Wu, R. (1982). Synthesis of a human insulin gene V. Enzymatic assembly, cloning and characterization of the human proinsulin DNA. *Gene* **17**, 279–289.

Burrell, C. J., Mackay, P., Greenaway, P. J., Hofschneider, P. H. and Murray, K. (1979). Expression in *E. coli* of hepatitis B virus DNA sequences cloned in plasmid pBR322. *Nature* **279**, 43–47.

Casadaban, M., Chou, J. and Cohen, S. N. (1980). *In vitro* gene fusions that join an enzymatically active β-galactosidase segment to amino-terminal fragments of exogenous proteins: *Escherichia coli* plasmid vectors for the detection and cloning of translational initiation signals. *J. Bact.* **143**, 971–980.

Chan, S. J., Weiss, J., Konrad, K., White, T., Bahl, C., Yu, S. D., Marks, D. and Steiner, D. F. (1981). Biosynthesis and periplasmic segregation of human proinsulin in *E. coli*. *Proc. Natn. Acad. Sci. U.S.A.* **78**, 5401–5405.

Chang, A. C. Y. and Cohen, S. N. (1974). Genome construction between bacterial species *in vitro*: Replication and expression of *Staphylococus* plasmid genes in *Escherichia coli. Proc. Natn. Acad. Sci. U.S.A.* **71,** 1030–1034.

Chang, A. C. Y., Lansman, R. A., Clayton, D. A. and Cohen, S. N. (1975). Studies of mouse mitochondrial DNA in *Escherichia coli*: Structure and function of the eukaryotic – prokaryotic chimeric plasmids. *Cell* **6,** 231–244.

Chang, A. C. Y., Nunberg, J. H., Kaufman, R. J., Erlich, H. A., Schimke, R. T. and Cohen, S. N. (1978). Phenotypic expression in *E. coli* of a DNA sequence coding for mouse dihydrofolate reductase. *Nature* **275,** 617–624.

Chang, A. C. Y., Erlich, H. A., Gunsalus, R. P., Nunberg, J. H., Kaufman, R. J., Schimke, R. T. and Cohen, S. N. (1980). Initiation of protein synthesis in bacteria at a translational start codon of mammalian cDNA: Effects of the preceding nucleotide sequence. *Proc. Natn. Acad. Sci. U.S.A.* **77,** 1442–1446.

Charnay, P., Perricaudet, M., Galibert, F. and Tiollais, P. (1978). Bacteriophage lambda and plasmid vectors allowing fusion of cloned genes in each of three translational phases. *Nucleic Acids Res.* **5,** 4479–4494.

Charnay, P., Louise, A., Fritsch, A., Perrin, D. and Tiollais, P. (1979). Bacteriophage lambda *Escherichia coli* K12 vector host system for gene cloning and expression under lactose promoter control II. DNA fragment insertion at the vicinity of the *lac* UV 5 promoter. *Molec. Gen. Genet.* **170,** 171–178.

Charnay, P., Gervais, M., Louise, A., Galibert, F. and Tiollais, P. (1980). Biosynthesis of hepatitis B virus surface antigen in *Escherichia coli. Nature* **286,** 893–895.

Chater, K. F., Hopwood, D. A., Kiester, T. and Thompson, C. J. (1982). Gene cloning in *Streptomyces. Curr. Topics Microbiol. Immunol.* **96,** 69–95.

Clarke, A. J. L., Adeniyi-Jones, R. O., Knight, G., Leiper, J. M., Wiles, P. G., Jones, R. H., Keen, H., Maccuish, A. C., Ward, J. D., Watkins, P. J., Cauldwell, J. M., Glynne, A. and Scotton, J. B. (1982). Biosynthetic human insulin in the treatment of diabetes. *Lancet* **2,** 354–357.

Cohen, B. J. and Richmond, J. E. (1982). Electron microscopy of hepatitis B core antigen synthesized in *E. coli. Nature* **296,** 677–678.

Corden, J., Wasylyk, B., Buchwalder, A., Sassone-Corsi, P., Kedinger, C. and Chambon, P. (1980). Promoter sequences of eukaryotic protein coding genes. *Science* **209,** 1406–1414.

Craig, R. and Hall, D. (1983). *In* "Genetic Engineering" (Williamson, R., ed). Vol. 4, pp. 57–125. Academic Press, London.

Crea, R., Kraszeuski, A., Hirose, T. and Itakura, K. (1978). Chemical synthesis of genes for human insulin. *Proc. Natn. Acad. Sci. U.S.A.* **75,** 5765–5769.

Davis, B. D. and Tai, P.-C. (1980). The mechanism of protein secretion across membranes. *Nature* **283,** 433–438.

Davis, A. R., Nayak, D. P., Cleda, M., Hiti, A. L., Dowbenko, D. and Kleid, D. G. (1981). Expression of antigenic determinants of the haemagglutinin gene of a human influenza virus in *E. coli. Proc. Natn. Acad. Sci. U.S.A.* **78,** 5376–5380.

de Boer, H. A., Comstock, L. J., Yansura, D. G. and Heynecker, H. L. (1982). Construction of a tandem *trp-lac* promoter for efficient and controlled expression of the human growth hormone gene in *Escherichia coli. In* "Promoter Structure and Function" (Rodriquez, R. L. and Chamberlin, M. J., eds) Praegar Publishers, New York.

De Maeyer, E., Skup, D., Prasad, K. S. N., De Maeyer-Guignard, J., Williams, B.,

Sharpe, G., Pioli, D., Hennam, J., Schuch, W. and Atherton, K. T. (1982). Expression of a chemically synthesized human α interferon gene. *Proc. Natn. Acad. Sci. U.S.A.* **79**, 4256–4259.

Derom, C., Gheysen, D. and Fiers, W. (1982). High level synthesis in *E. coli* of the SV40 small-t antigen under control of the bacteriophage lambda P_L promoter. *Gene* **17**, 45–54.

Derynck, R., Remaut, E., Saman, E., Stanssens, P., De Clercq, E., Content, T. and Fiers, W. (1980). Expression of human fibroblast interferon gene in *E. coli. Nature* **287**, 193–197.

Edens, L., Heslinga, L., Klok, R., Ledeboer, A. M., Maat, J., Toonen, M. Y., Visser, C. and Verrips, C. T. (1982). Cloning of cDNA encoding the sweet-tasting plant protein thaumatin and its expression in *Escherichia coli. Gene* **18**, 1–12.

Edge, M. D., Greene, A. R., Heathcliffe, G. R., Meacock, P. A., Schuch, W., Scanlon, D. B., Atkinson, T. C., Newton, C. R. and Markham, A. F. (1981). Total synthesis of a human leukocyte interferon gene. *Nature* **191**, 756–762.

Edman, J. C., Hallewell, R. A., Valenzuela, P., Goodman H. M. and Rutter, W. J. (1981). Expression of hepatitis B surface and core antigens in *E. coli. Nature* **291**, 503–506.

Emtage, J. S., Tacon, W. C. A., Catlin, G. H., Jenkins, B., Porter, A. G. and Carey, N. H. (1980). Influenza antigenic determinants are expressed from haemagglutinin genes cloned in *E. coli. Nature* **283**, 171–174.

Emtage, J. S., Angal, S., Doel, M. T., Harris, T. J. R., Jenkins, B., Lilley, G. and Lowe, P. A. (1983). In preparation.

Erwin, C. R., Maurer, R. A. and Donelson, J. E. (1980). A bacterial cell that synthesises a protein containing the antigenic determinants of rat prolactin. *Nucleic Acids Res.* **8**, 2537–3546.

Fraser, T. H. and Bruce, B. J. (1978). Chicken ovalbumin is synthesised and secreted by *E. coli. Proc. Natn. Acad. Sci. U.S.A.* **75**, 5936–5940.

Fuller, F. (1982). A family of cloning vectors containing the *lac* UV5 promoter. *Gene* **19**, 43–54.

Garapin, A. C., Colbere-Garapin, F., Cohen-Solal, M., Horodniceanu, F. and Kourilsky, P. (1981). Expression of herpes simplex virus type 1 thymidine kinase gene in *Escherichia coli. Proc. Natn. Acad. Sci. U.S.A.* **78**, 815–819.

Gatenby, A. A., Castleton, J. A. and Saul, M. W. (1981). Expression in *E. coli* of maize and wheat chloroplast genes for large subunit of ribulose bisphosphate carboxylase. *Nature* **291**, 117–121.

Gelfand, D. H., Shepard, H. M., O'Farrell, P. H. and Polisky, B. (1978). Isolation and characterisation of a Col E1 derived plasmid copy-number mutant. *Proc. Natn. Acad. Sci. U.S.A.* **75**, 55869–55873.

Gentz, R., Lagner, A., Chang, A. C. Y., Cohen, S. N. and Bujard, H. (1981). Cloning and analysis of strong promoters is made possible by the downstream placement of an RNA termination signal. *Proc. Natn. Acad. Sci. U.S.A.* **78**, 4936–4940.

Gheysen, D., Iserentant, D., Derom, C. and Fiers, W. (1982). Systematic alteration of the nucleotide sequence preceding the translation initiation codon and the effects on bacterial expression of the cloned SV40 small-t antigen gene. *Gene* **17**, 55–63.

Gilmer, T. M. and Erikson, R. L. (1981). Rous sarcoma virus transforming protein p60 src expressed in *E. coli* functions as a protein kinase. *Nature* **294**, 771–773.

Gilmer, T. M., Parsons, J. T. and Erikson, R. L. (1982). Construction of plasmids for expression of Rous sarcoma virus transforming protein p60 src in *Escherichia coli. Proc. Natn. Acad. Sci. U.S.A.* **79**, 2152–2156.

Goeddel, D. V., Kleid, D. G., Bolivar, F., Heynecker, H. C., Yansura, D. G., Crea, R., Hirose, T., Kraszeuski, A., Itakura, K. and Riggs, A. D. (1979a). Expression in *E. coli* of chemically synthesised genes for human insulin. *Proc. Natn. Acad. Sci. U.S.A.* **76**, 106–110.

Goeddel, D. V., Heynecker, H. L., Hozumi, T., Arentzen, R., Itakura, K., Yansura, D. G., Ross, M. J., Miozzari, G., Crea, R. and Seeburg, P. H. (1979b). Direct expression in *Escherichia coli* of a DNA sequence coding for human growth hormone. *Nature* **281**, 544–548.

Goeddel, D. V., Yelverton, E., Ullrich, A., Heynecker, H. L., Miozzari, G., Holmes, W., Seeburg, P. H., Dull, T., May, L., Stebbing, N., Crea, R., Maeda, S., McCandliss, N., Sloma, A., Tabar, J. M., Gross, M., Familletti, P. C. and Pestka, S. (1980a). Human leukocyte interferon production in *E. coli* is biologically active. *Nature* **287**, 411–416.

Goeddel, D. V., Shepard, H. M., Yelverton, E., Leung, D. and Crea, R. (1980b). Synthesis of human fibroblast interferon by *E. coli. Nucleic Acids Res.* **8**, 4057–4074.

Goeddel, D. V., Leung, D. W., Dull, T. J., Gross, M., Lawn, R. M., McCandliss, R., Seeburg, P. H., Ullrich, A., Yelverton, E. and Gray, P. W. (1981). The structure of eight distinct cloned human leukocyte interferon cDNAs. *Nature* **290**, 20–26.

Goldberg, A. L., Swamy, K. H. S., Chung, C. H. and Larimore, F. S. (1980). Proteases in *Escherichia coli. Methods Enzymol.* **80**, 680–702.

Gottesman, S., Gottesman, M., Shaw, J. E. and Pearson, M. L. (1981). Protein degradation in *E. coli*: The Lon mutation and bacteriophage lambda N and CII protein stability. *Cell* **24**, 225–233.

Grantham, R., Gautier, C., Gouy, M., Jacobzone, M. and Mercier, R., (1981). Codon catalog usage is a genome strategy modulated for gene expressivity. *Nucleic Acids Res.* **9**, 443–r75.

Gray, P. W., Leung, D. W., Pennica, D., Yelverton, E., Wajarian, R., Simonsen, C. C., Derynck, R., Sherwood, P. J., Wallace, D. M., Berger, S. L., Levinson, A. D. and Goeddel, D. V. (1982). Expression of human immune interferon cDNA in *E. coli* and monkey cells. *Nature* **295**, 503–508.

Greenblatt, J. (1981). Regulation of transcription termination by the N gene protein of bacteriophage λ. *Cell* **24**, 8–9.

Grosjean, H. and Fiers, W. (1982). Preferential codon usage in prokaryotic genes. The optimal codon-anticodon interaction energy and selective codon usage in efficiently expressed genes. *Gene* **18**, 199–209.

Gryczan, T. J. (1981). Molecualr cloning in *Bacillus subtilis. In* "The Molecular Biology of the Bacilli" (Dubnau, D., ed), Vol. 1, pp. 307–329. Academic Press, New York.

Guarente, L., Lauer, G., Roberts, T. M. and Ptashne, M. (1980). Improved methods for maximizing expressing of a cloned gene: A bacterium that synthesises rabbit β-globin. *Cell* **20**, 543–553.

Hallewell, R. A. and Emtage, S. (1980). Plasmid vectors containing the tryptophan operon promoter suitable for efficient regulated expression of foreign genes. *Gene* **9**, 24–47.

Hardy, K., Stahl, S. and Kupper, H. (1981). Production in *B. subtilis* of hepatitis B core antigen and of major antigen of foot and mouth disease virus. *Nature* **293**, 481–483.

Hautala, J. A., Bassett, C. L., Giles, N. H. and Kushner, S. R. (1979). Increased expression of a eukaryotic gene in *E. coli* through stabilization of its messenger RNA. *Proc. Natn. Acad. Sci. U.S.A.* **76**, 5774–5778.

Heiland, I. and Gething, M. J. (1981). Cloned copy of the haemagglutinin gene codes for human influenza antigenic determinants in *E. coli. Nature* **292**, 851–852.

Hennam, J. F., Cunningham, A. E., Sharpe, G. S. and Atherton, K. T. (1982). Expression of eukaryotic coding sequences in *Methylophilus methylotrophus. Nature* **297**, 80–82.

Hershfield, M. V., Boyer, H. W., Yanofsky, C., Lovett, M. and Helinski, D. R. (1974). Plasmid Col E1 as a molecular vehicle for cloning and amplification of DNA. *Proc. Natn. Acad. Sci. U.S.A.* **71**, 3455–3459.

Hinnen, A. and Meyhack, B. (1982). Vectors for cloning in yeast. *Curr. Topics Microbiol. Immunol.* **96**, 101–117.

Hintz, R. L., Wilson, D. M., Finno, J., Rosenfeld, R. G., Bennett, A. and McClellan, B. (1982). Biosynthetic methionyl human growth hormone is biologically active in adult man. *Lancet* **1**, 1276–1279.

Hintzeman, R. A., Hagie, F. E., Levine, H. L., Goeddel, D. V., Ammerer, G. and Hall, B. D. (1981). Expression of a human gene for interferon in yeast. *Nature* **293**, 717–722.

Hollenberg, C. P. (1982). Cloning with 2 μm DNA vectors and the expression of foreign genes in *Saccharomyces cerevisiae. Curr. Topics Microbiol. Immunol.* **96**, 119–144.

Hopkins, A. S., Murray, N. E. and Brammar, W. J. (1976). Characterisation of λ *trp*-transducing bacteriophages made *in vitro. J. Mol Biol.* **107**, 549–569.

Horwich, A., Koop, A. H. and Eckhart, W. (1980). Expression of a gene for the polyoma small t antigen in *Escherichia coli. J. Virol.* **36**, 125–132.

Ikemura, T. (1981). Correlation between the abundance of *E. coli* transfer RNAs and the occurrence of respective codons in its protein genes. *J. Mol. Biol.* **146**, 1–21.

Iserentant, D. and Fiers, W. (1980). Secondary structure of mRNA and efficiency of translation initiation. *Gene* **9**, 1–12.

Itakura, K., Hirose, T., Crea, R., Riggs, A. D., Heyneker, H. L., Bolivar, F. and Boyer, H. W. (1977). Expression in *E. coli* of a chemically synthesised gene for the hormone somatostatin. *Science* **198**, 1056–1063.

Jay, G., Khoury, G., Seth, A. K. and Jay, E. (1981). Construction of a general vector for efficient expression of mammalian proteins in bacteria: Uses of a synthetic ribosome binding site. *Proc. Natn. Acad. Sci. U.S.A.* **78**, 5543–5548.

Johnson, A. D., Poteete, A. R., Lauer, G., Sauer, R. T., Ackers, G. and Ptashne, M. (1981). λ repressor and cro—components of an efficient molecular switch. *Nature* **294**, 217–223.

Kedes, L. H., Chang, A. C. Y., Houseman, D. and Cohen, S. N. (1975). Isolation of histone genes from unfractionated sea urchin DNA by sub culture and cloning in *E. coli. Nature* **255**, 533–538.

Keefer, L. M., Piron, M. A. and De Meyts, P. (1981). Human insulin prepared by recombinant DNA techniques and native human insulin interact identically with insulin receptors. *Proc. Natn. Acad. Sci. U.S.A.* **78**, 1391–1395.

Kelley, W. S., Chalmers, K. and Murray, N. E. (1977). Isolation and characterisation of a λ *pol* A transducing phage. *Proc. Natn. Acad. Sci. U.S.A.* **74**, 5362–5366.

Keshet, E., Rosner, A., Bernstein, Y., Gorecki, M. and Aviv, H. (1981). Cloning of bovine growth hormone gene and its expression in bacteria. *Nucleic Acids Res.* **9**, 19–30.

Kit, S., Otsuka, H., Qavi, H. and Kit, M. (1981). Functional expression of the herpes simplex virus thymidine kinase gene in *E. coli* K-12. *Gene* **16**, 287–295.

Kleid, P. G., Yansura, D., Small, B., Dowbenko, D., Moore, D., Grubman, M., McKercher, P., Morgan, D. O., Robertson, B. H. and Bachrach, H. L. (1981a). Cloned viral protein vaccine for foot and mouth disease: Responses in cattle and swine. *Science* **214**, 1125–1129.

Kleid, D. G., Yansura, D. G., Heynecker, H. I. and Miozzari, G. F. (1981b). A method of producing a polypeptide product and a plasmidic expression vehicle therefore, a method of creating an expression plasmid, a method of cleaving double stranded DNA and specific plasmids. *UK Patent Application 2073203 A.*

Kohr, W. J., Keck, R. and Harkins, R. N. (1982). Characterization of intact and trypsin-digested biosynthetic human growth hormone by high-pressure liquid chromatography. *Anal. Biochem.* **122**, 348–359.

Kourilsky, P., Gros, D., Rougeon, F. and Mach, B. (1977). Transcription of a mammalian sequence under phage λ control. *Nature* **267**, 637–639.

Kozak, M. (1981). Possible role of flanking nucleotides in recognition of the AUG initiator codon by eukaryotic ribosomes. *Nucleic Acids Res.* **9**, 5233–5252.

Kupper, H., Keller, W., Kurz, C., Forss, S., Schaller, H., Franze, R., Strohmaier, K., Marquardt, O., Zaslavsky, V. G. and Hofschneider, P. H. (1981). Cloning of cDNA of major antigen of foot and mouth disease virus and expression in *E. coli. Nature* **289**, 555–559.

Lathe, R. F., Lecocq, J. P. and Everett, D. (1983). *In* "Genetic Engineering" (Williamson, R., ed) Vol. 4, pp. 1–56. Academic Press, New York and London.

Lauer, G., Pastrana, R., Sherley, J. and Ptashne, M. (1981). Construction of overproducers of bacteriophage 434 repressor and cro proteins. *J. Mol. Appl. Genet.* **1**, 139–147.

Lawn, R., Adelman, J., Bock, S. C., Franke, A. E., Houck, C. M., Najarian, R., Seeburg, P. H. and Wion, K. L. (1981). The sequence of human serum albumin cDNA and its expression in *E. coli. Nucleic Acids Res.* **9**, 6103–6114.

Mackay, P., Pasek, M., Magazin, M., Kovakic, R. T., Allet, B., Stahl, S., Gilbert, W., Schaller, H., Bruce, S. A. and Murray, K. (1981). Production of immunologically active surface antigens of hepatitis B virus by *Escherichia coli. Proc. Natn. Acad. Sci. U.S.A.* **78**, 4510–4514.

Martial, J. A., Hallewell, R. A., Baxter, J. D. and Goodman, H. M. (1979). Human growth hormone: Complementary DNA cloning and expression in bacteria. *Science* **205**, 602–607.

McAllister, W. T., Morris, C., Rosenberg, A. H. and Studier, F. W. (1981). Utilization of bacteriophage T7 late promoters in recombinant plasmids during infection. *J. Mol. Biol.* **153**, 527–544.

McGrath, J. P. and Levinson, A. D. (1982). Bacterial expression of an enzymatically active protein encoded by RSV *src* gene. *Nature* **295**, 423–435.

Meacock, P. A. and Cohen, S. N. (1980). Partitioning of bacterial plasmids during cell division: A cis-acting locus that accomplishes stable plasmid inheritance. *Cell* **20**, 529–540.

Mercereau-Puijalon, O., Royal, A., Carni, B., Garapin, A., Krust, A., Gannon, F.

and Kourilsky, P. (1978). Synthesis of an ovalbumin-like protein by *E. coli* K12 harbouring a recombinant plasmid. *Nature* **275**, 505–510.

Mercereau-Puijalon, O., Lacroute, F. and Kourilsky, P. (1980). Synthesis of a chicken ovalbumin-like protein in the yeast *Saccharomyces cerevisiae*. *Gene* **11**, 163–167.

Miller, J. H. and Reznikoff, W. S. (1980). The Operon. Cold Spring Harbor Laboratory.

Moir, A. and Brammer, W. J. (1976). The use of specialised transducing phages in the amplification of enzyme production. *Molec. Gen. Genet.* **149**, 87–99.

Morrow, J. F., Cohen, S. N., Chang, A. C. Y., Boyer, H. W., Goodman, H. M. and Helling, R. B. (1974). Replication and transcription of eukaryotic DNA in *Escherichia coli*. *Proc. Natn. Acad. Sci. U.S.A.* **71**, 1743–1747.

Mory, Y., Chernajovsky, Y., Feinstein, S. I., Chen, L., Nir, U., Weissenbach, J., Malpiece, Y., Tiollais, P., Marks, D., Ladner, M., Colby, C. and Revel, M. (1981). Synthesis of human interferon β in *Escherichia coli* infected by a λ phage recombinant containing a human genomic fragment. *Eur. J. Biochem.* **120**, 197–202.

Muesing, M., Tamm, J., Shepard, H. M. and Polisky, B. (1981). A single base pair alteration is responsible for the DNA overproduction phenotype of a plasmid copy number mutant. *Cell* **24**, 235–242.

Murray, N. E. and Kelley, W. S. (1979). Characterisation of λ*pol* A transducing phages; effective expression of the *E. coli* pol A gene. *Molec. Gen. Genet.* **175**, 77–87.

Murray, N. E., Bruce, S. A. and Murray, K. (1979). Molecular cloning of the DNA ligase gene from bacteriophage T4. *J. Mol. Biol.* **132**, 493–505.

Nagata, S., Taira, H., Hall, A., Johnsrud, L., Streuli, M., Ecsodi, J., Boll, W., Cantell, K. and Weissmann, C. (1980a). Synthesis in *E. coli* of a polypeptide with human leukocyte interferon activity. *Nature* **284**, 316–320.

Nagata, S., Mantei, N. and Weissmann, C. (1980b). The structure of one of the eight or more distinct chromosomal genes for human interferon-α. *Nature* **287**, 401–408.

Nakamura, K. and Inouye, M. (1982). Construction of versatile expression cloning vehicles using the lipoprotein gene of *E. coli*. *EMBO Journal* **1**, 771–775.

Nakamura, K., Masui, Y. and Inouye, M. (1982). Use of a *lac* promoter-operator fragment as a transcriptional control switch for expression of the constitutive *lpp* gene in *Escherichia coli*. *J. Mol. Appl. Genet.* **1**, 289–299.

Olson, K. C., Fenno, J., Lin, N., Harkins, R. N., Snider, C., Kohr, W., Ross, M. J., Fodge, D., Prender, G. and Stebbing, N. (1981). Purified human growth hormone from *E. coli* is biologically active. *Nature* **293**, 408–411.

O'Farrell, P. H., Polisky, B. and Gelfand, D. H. (1978). Regulated expression by readthrough translation from a plasmid-encoded β-galactosidase. *J. Bact.* **134**, 645–654.

Panasenko, S. M., Cameron, J. R., Davis, R. W. and Lehman, I. R. (1977). Five-hundredfold overproduction of DNA ligase after induction of a hybrid λ lysogen constructed *in vitro*. *Science* **196**, 188–189.

Pasek, M., Goto, T., Gilbert, W., Zink, B., Schaller, H., Mackay, P., Leadbetter, G. and Murray, K. (1979). Hepatitis B virus genes and their expression in *E. coli*. *Nature* **282**, 575–579.

Polisky, B., Bishop, R. J. and Gelfand, D. H. (1976). A plasmid cloning vehicle allowing regulated expression of eukaryotic DNA in bacteria. *Proc. Natn. Acad. Sci. U.S.A.* **73**, 3900–3904.

Rao, R. and Rogers, S. G. (1978). A thermoinducible λ phage-Col E1 plasmid chimera for the overproduction of gene products from cloned DNA segments. *Gene* **3**, 247–263.

Ratzkin, B. and Carbon, J. (1977). Functional expression of cloned yeast DNA in *Escherichia coli*. *Proc. Natn. Acad. Sci. U.S.A.* **74**, 487–491.

Ratzkin, B., Lee, S. G., Schrenk, W. J., Roychoudhury, R., Chen, M., Hamilton, T. A. and Hung, P. P. (1981). Expression in *Escherichia coli* of biologically active enzyme by a DNA sequence coding for the human plasminogen activator urokinase. *Proc. Natn. Acad. Sci. U.S.A.* **78**, 3313–3317.

Remaut, E., Stanssens, P. and Fiers, W. (1981). Plasmid vectors for high-efficiency expression controlled by the P_L promoter of coliphage lambda. *Gene* **15**, 81–93.

Rigby, P. W. J. (1982). Expression of cloned genes in eukaryotic cells using vector systems derived from viral replicons. *In* "Genetic Engineering" (Williamson, R., ed), Vol. 3, pp. 88–141. Academic Press, London.

Roberts, J. L., Seeburg, P. H., Shine, J., Herbert, E., Baxter, J. D. and Goodman, H. M. (1979a). Corticotropin and β endorphin: Construction and analysis of recombinant DNA complementary to mRNA for the common precursor. *Proc. Natn. Acad. Sci. U.S.A.* **76**, 2153–2157.

Roberts, T. M., Kacich, R. and Ptashne, M. (1979b). A general method for maximizing the expression of a cloned gene. *Proc. Natn. Acad. Sci. U.S.A.* **76**, 760–764.

Roberts, T. M., Bikel, I., Rogers-Yocum, R., Livingston, D. M. and Ptashne, M. (1979c). Synthesis of simian virus 40 t antigen in *E. coli*. *Proc. Natn. Acad. Sci. U.S.A.* **76**, 5596–5600.

Rose, J. K. and Shafferman, A. (1981). Conditional expression of the vesicular stomatitis virus glycoprotein gene in *E. coli*. *Proc. Natn. Acad. Sci. U.S.A.* **78**, 6670–6674.

Rosenberg, M. and Court, D. (1979). Regulatory sequences involved in the promotion and termination of RNA transcription. *Ann. Rev. Genetics* **13**, 319–353.

Rosenfeld, R. G., Aggarwal, B. B., Hintz, R. L. and Dollar, L. A. (1982). Recombinant DNA-derived methionyl human growth hormone is similar in membrane binding properties to human pituitary growth hormone. *Biochem. Biophys. Res. Comm.* **106**, 202–209.

Scott, G. M. and Tyrrell, D. A. J. (1980). Interferon: Therapeutic fact or fiction for the '80s? *Brit. Med. J.* **280**, 1558–1562.

Seeburg, P. H., Shine, J., Martial, J. A., Ivarie, R. D., Morris, J. A., Ullrich, A., Baxter, J. D. and Goodman, H. M. (1978). Synthesis of growth hormone by bacteria. *Nature* **276**, 795–598.

Shemyakin, M. F., Chestukin, A. V., Dolganov, G. M., Khodkova, E. M., Monastyrskaya, G. S. and Sverdlov, E. D. (1980). Leu-enkephalin purification from *E. coli* cells carrying the plasmid with fused synthetic leu-enkephalin gene. *Nucleic Acids Res.* **8**, 6163–6173.

Shepard, H. M., Yelverton, E. and Goeddel, D. V. (1982). Increased synthesis in *E. coli* of fibroblast and leukocyte interferons through alterations in ribosome binding sites. *DNA* **1**, 125–131.

Shimatake, H. and Rosenberg, M. (1981). Purified λ regulatory protein CII positively activates promoters for lysogenic development. *Nature* **292**, 128–132.

Shine, J. and Dalgarno, L. (1975). Determinant of cistron specificity in bacterial ribosomes. *Nature* **254**, 34–38.

Shine, J., Fettes, I., Lan, N. C. Y., Roberts, J. L. and Baxter, J. D. (1980).

Expression of a cloned β-endorphin gene sequence by *Escherichia coli. Nature* **285**, 456–461.
Shortle, D., Dimaio, D. and Nathans, D. (1981). Directed mutagenesis. *Ann. Rev. Genetics* **15**, 265–294.
Siebenlist, U., Simpson, R. B. and Gilbert, W. (1980). *E. coli* polymerase interacts homologously with two different promoters. *Cell* **20**, 269–281.
Slocombe, P., Easton, A., Boseley, P. and Burke, D. C. (1982). High level expression of an interferon $\alpha 2$ gene cloned in phage M13 mp7 and subsequent purification with a monoclonal antibody. *Proc. Natn. Acad. Sci. U.S.A.* **79**, 5455–5459.
Smith, M. and Gillam, S. (1981). Constructed mutants using synthetic oligodeoxyribonucleotides as site-specific mutagens. *In* "Genetic Engineering" (Setlow, J. K. and Hollaender, A., eds) Vol. 3, pp. 1–32. Plenum Press, New York.
Sninsky, J. J., Uhlin, B. E., Gustafsson, P. and Cohen, S. N. (1981). Construction and characterisation of a novel two-plasmid system for accomplishing temperature-regulated amplified expression of cloned adventitious genes in *Escherichia coli. Gene* **16**, 275–286.
Stahl, S., Mackay, P., Magazin, M., Bruce, S. A. and Murray, K. (1982). Hepatitis B virus core antigen: Synthesis in *E. coli* and application in diagnosis. *Proc. Natn. Acad. Sci. U.S.A.* **79**, 1606–1610.
Steitz, J. A. (1979). Genetic signals and nucleotide sequences in messenger RNA. *In* "Biological Regulation and development 1. Gene Expression". (Goldberger, R. F., ed) pp. 349–399. Plenum Press, New York.
Streuli, M., Nagata, S. and Weissman, C. (1980). At least three human type α interferons: Structure of α_2. *Science* **209**, 1343–1347.
Streuli, M., Hall, A., Boll, W., Stewart, W. E., Nagata, S. and Weissmann, C. (1981). Target cell specificity of two species of human interferon-α produced in *E. coli* and of hybrid molecules derived from them. *Proc. Natn. Acad. Sci. U.S.A.* **78**, 2848–2852.
Stewart, W. E. II, Sarkar, F. H., Taira, H., Hall, A.. Nagata, S. and Weissmann, C. (1980). Comparisons of several biological and physiochemical properties of human leukocyte interferons produced by human leukocytes and by *E. coli. Gene* **11**, 181–186.
Struhl, K., Cameron, J. R. and Davis, R. W. (1976). Functional genetic expression of eukaryotic DNA in *Escherichia coli. Proc. Natn. Acad. Sci. U.S.A.* **73**, 1471–1475.
Struhl, K. and Davis, R. W. (1977). Production of a functional eukaryotic enzyme in *Escherichia coli*: Cloning and expression of the yeast structural gene for imidazole-glycerol phosphate dehydratase (his 3). *Proc. Natn. Acad. Sci. U.S.A.* **74**, 5255–5259.
Sung, W. L., Hsiung, H. M., Brousseau, R., Michniewicz, J., Wu, R. and Narang, S. A. (1979). Synthesis of the human insulin gene. Part II. Further improvements in the modified phosphotriester method and the synthesis of seventeen deoxyribooligonucleotide fragments constituting human insulin chains B and mini-C DNA. *Nucleic Acids Res.* **7**, 2199–2212.
Sutcliffe, J. G. (1979). Complete nucleotide sequence of the *Escherichia coli* plasmid pBR322. *Cold Spring Harbor Symp. Quant. Biol.* **43**, 77–90.
Suzuki, M., Sumi, S.-I., Hasegawa, A., Nishizawa, T., Miyoshi, K.-I., Wakisaka, S., Miyake, T. and Misoka, F. (1982). Production in *Escherichia coli* of biologically active secretin a gastro-intestinal hormone. *Proc. Natn. Acad. Sci. U.S.A.* **79**, 2475–2479.

Szybalski, E. H. and Szybalski, W. (1979). A comprehensive molecular map of bacteriophage λ. *Gene* **7**, 217–270.

Tacon, W., Carey, N. H. and Emtage, S. (1980). The construction and characterization of plasmid vectors suitable for the expression of all DNA phases under the control of the *E. coli* tryptophan promoter. *Mol. Gen. Genet.* **177**, 427–438.

Talmadge, K. and Gilbert, W. (1982). Cellular location affects protein stability in *Escherichia coli. Proc. Natn. Acad. Sci. U.S.A.* **79**, 1830–1833.

Talmadge, K., Kaufman, J. and Gilbert, W. (1980a). Bacteria mature preproinsulin to proinsulin. *Proc. Natn. Acad. Sci. U.S.A.* **77**, 3988–3992.

Talmadge, K., Stahl, S. and Gilbert, W. (1980b). Eukaryotic signal sequence transports insulin antigen in *E. coli. Proc. Natn. Acad. Sci. U.S.A.* **77**, 3369–3373.

Talmadge, K., Brosius, J. and Gilbert, W. (1981). An internal signal sequence directs secretion and processing of proinsulin in bacteria. *Nature* **294**, 176–178.

Tanaka, S., Oshima, T., Ohsue, K., Ono, T., Oikawa, S., Takano, I., Noguchi, T., Kangawa, K., Minamino, N. and Matsuo, H. (1982). Expression in *Escherichia coli* of chemically synthesized gene for a novel opiate peptide α neo-endorphin. *Nucleic Acids Res.* **10**, 1741–1754.

Taniguchi, T., Guarente, L., Roberts, T. M., Kimelman, D., Douhan, J. and Ptashne, M. (1980). Expression of the human fibroblast interferon gene in *E. coli. Proc. Natn. Acad. Sci. U.S.A.* **77**, 5230–5233.

Thummel, C. S., Burgess, T. L. and Tjian, R. (1981). Properties of simian virus 40 small t antigen overproduced in bacteria. *J. Virol.* **37**, 683–697.

Tuite, M. F., Dobson, M. J., Roberts, N. A., King, R. M., Burke, D. C., Kingsman, S. M. and Kingsman, A. J. (1982). Regulated high efficiency expression of human interferonalpha in *Saccharomyces cerevisiae. EMBO Journal* 1, 603–608.

Twigg, A. and Sherratt, D. (1980). Trans complementable copy number mutants of plasmid Col E1. *Nature* **283**, 216–218.

Uhlin, B. E., Molin, S., Gustafsson, P. and Nordstrom, K. (1979). Plasmids with temperature-dependent copy number for amplification of cloned genes and their products. *Gene* **6**, 91–106.

Vapnek, D., Hautala, J., Jacobson, J. W., Giles, N. H. and Kushner, S. R. (1977). Expression in *Escherichia coli* K12 of the structural gene for catabolic dehydroquinase of *Neurospora crassa. Proc. Natn. Acad. Sci. U.S.A.* **74**, 3508–3512.

Villa-Komaroff, L., Efstratiadis, A., Broome, S., Lomedico, P., Tizard, R., Naber, S. P., Chick, W. L. and Gilbert, W. (1978). A bacterial clone synthesising proinsulin. *Proc. Natn. Acad. Sci. U.S.A.* **75**, 3727–3731.

Weck, P.K., Apperson, S., Stebbing, N., Gray, P. W., Leung, D., Shepard, H. M. and Goeddel, D. V. (1981). Antiviral activities of hybrids of two major human leukocyte interferons. *Nucleic Acids Res.* **9**, 6153–6166.

Wetzel, R. (1981). Assignment of the disulphide bonds of leukocyte interferon. *Nature* **289**, 606–607.

Wetzel, R., Heyneker, H. L., Goeddel, D. V., Thurman, G. B. and Goldstein, A. L. (1980). Production of biologically active Nα-desacetylthymosin α_1 in *E. coli* through expression of a chemically synthesized gene. *Biochemistry* **19**, 6096–6104.

Wetzel, R., Kleid, D. G., Crea, R., Heyneker, H. L., Yansura, D. G., Hirose, T., Kraszeuski, A., Riggs, A. D., Itakura, K. and Goeddel, D. V. (1981a). Expression in *E. coli* of a chemically synthesized gene for a mini C analog of human proinsulin. *Gene* **16**, 63–71.

Wetzel, R., Perry, L. J., Estell, D. A., Lin, N., Levine, H. L., Slinker, B., Fields, F., Ross, M. J. and Shively, J. (1981b). Properties of a human alpha-interferon purified from *E. coli* extracts. *J. Interferon Res.* **1**, 381–390.

Yanofsky, C. (1981). Attenuation in the control of expression of bacterial operons. *Nature* **289**, 751–758.

Yanofsky, C., Platt, T., Crawford, I. P., Nichols, B. P., Christie, G. E., Horowitz, H., Van Cleemput, M. and Wu, A. M. (1981). The complete nucleotide sequence of the tryptophan operon of *E. coli. Nucleic Acids Res*, **9**, 6647–6668.

Yelverton, E., Leung, D., Weck, P., Gray, P. W. and Goeddel, D. V. (1981). Bacterial synthesis of a novel human leukocyte interferon. *Nucleic Acids Res*, **9**, 731–741.